FOOD AND BEVERAGE CONSUMPTION AND HEALTH

SEAFOOD FRAUD AND SAFETY: BACKGROUND AND ISSUES

FOOD AND BEVERAGE CONSUMPTION AND HEALTH

Additional books in this series can be found on Nova's website under the Series tab.

Additional E-books in this series can be found on Nova's website under the E-books tab.

FOOD AND BEVERAGE CONSUMPTION AND HEALTH

SEAFOOD FRAUD AND SAFETY: BACKGROUND AND ISSUES

KAITLYN O. HARRIS
EDITOR

Nova Science Publishers, Inc.
New York

For permission to use material from this book please contact us:
Telephone 631-231-7269; Fax 631-231-8175
Web Site: http://www.novapublishers.com

LIBRARY OF CONGRESS CATALOGING-IN-PUBLICATION DATA

Seafood fraud and safety : background and issues / editor: Kaitlyn O.
Harris.
p. cm.
Includes index.
ISBN 978-1-61761-500-9 (hardcover)
1. Seafood industry--Corrupt practices--United States. 2. Fraud--United
States--Prevention. 3. Seafood--United States--Safety measures. I. Harris,
Kaitlyn O.
HD9455.S42 2011
363.19'29--dc22
2010031751

Published by Nova Science Publishers, Inc. † New York

CONTENTS

PREFACE

Media attention has focused on recent incidents of fraud relating to seafood where restaurants knowingly serve lower-priced fish than identified on menus; lower-priced species are marketed commercially as higher-priced species; packaged weights of seafood are less than labeled weights, and extra water is added to seafood to increase the total product weight; all of which raise public concern. Most seafood fraud is based on supplying the consumer with something different from and inferior to the product expected. This book explores the questions Congress faces of whether the laws applicable to fraudulent seafood sales and marketing are clear and enforceable, whether agency enforcement efforts targeting seafood fraud are adequate, and whether penalties for seafood fraud are a deterrent

Chapter 1- Congress is facing questions of whether the law applicable to fraudulent seafood sales and marketing is clear and enforceable, whether agency enforcement efforts targeting seafood fraud are adequate, and whether the penalties for seafood fraud are a deterrent. Congress may become involved in oversight of how federal agencies are addressing these issues, and legislation related to these concerns may be considered.

With increased seafood imports and decreased monitoring, fraud and deception in seafood marketing is becoming more widespread. The flesh of many fish species is similar in taste and texture and, therefore, it is difficult to identify species in fillet form, especially after preparation for consumption. Thus, it can be relatively easy to substitute an inexpensive species for one of higher value. Inaccurate (low) counts or net weights ("short weighting") result in consumers receiving less for their money than advertised and anticipated. Overbreading may cause consumers to pay shrimp prices for excess bread crumbs. Excessive amounts of glaze (overglazing) can deliberately be used to increase the apparent weight, and therefore the apparent value, of the delivered product. In addition, some new treatment procedures by the seafood industry, such as carbon monoxide/tasteless smoke, are being questioned for their potential to deceive consumers. Since food safety and bioterrorism concerns have taken precedence, regulatory agencies have been less able to maintain control of economic fraud.

Chapter 2- In 2007, Americans consumed almost 5 billion pounds of seafood. Most seafood buyers, at many levels—importers, distributors, supermarkets, restaurants, and individual consumers—assume that the seafood they buy is what the seller claims it is. However, this is not always the case. Sometimes seafood products are mislabeled for financial gain—an activity called seafood fraud. Three federal agencies play key roles in detecting and preventing seafood fraud: the Department of Homeland Security's Customs and Border Protection (CBP), the Department of Commerce's National Marine Fisheries Service (NMFS), and the Department of Health and Human Services' Food and Drug Administration (FDA). GAO was asked to determine (1) the actions key federal agencies take to help detect and prevent seafood fraud and (2) the extent to which these key federal agencies collaborate with each other to help detect and prevent seafood fraud. GAO reviewed data and documents from each agency on actions to detect and prevent seafood fraud, and interviewed agency officials and other key stakeholders.

Chapter 3- On May 28, 2009 a Forum was held at the National Institute of Standards and Technology (NIST), in Gaithersburg Maryland (agenda attached) to discuss the problem and solutions to short net weight fraud occurring in the Seafood industry. The meeting was a collaborative effort attended by 30 participants representing Federal Agencies from Customs and Border Protection (CBP),Food and Drug Administration (FDA), National Marine Fisheries Service (NMFS), National Institute of Standards and Technology (NIST), and the U.S. Department of Agriculture (USDA); State Weights and Measures Officials from California, Florida, Maryland, Minnesota (teleconference), New York, Virginia, and Wisconsin; Trade Associations from American Frozen Food Institute (AFFI), Better Seafood Board (BSB), International Food Distributors Association (I FDA), National Association of Attorney Generals (NAAG), and National Fisheries Institute (NFI). Also participating were industry representatives from Maritime Products International, Pilot Brands, and Shinnecock Ocean Seafood - A Native American Company.

Chapter 4- Although seafood consumption can contribute to a healthy diet, some fish and shellfish can cause foodborne illnesses or contain environmental contaminants. Are current food safety programs sufficiently protecting consumers, and if not, what changes should be considered? A complicating factor is that most of the seafood consumed in the United States is from imports.

The Food and Drug Administration (FDA) within the Department of Health and Human Services plays the lead role in ensuring the safety of both domestic and imported fish and shellfish, but other agencies, including the National Marine Fisheries Service in the Department of Commerce and the Food Safety and Inspection Service in the U.S. Department of Agriculture, also have notable responsibilities.

In the 111th Congress, the food safety bills specific to seafood include S. 92, aimed at violative seafood imports, and H.R. 1370, authorizing $15 million annually to strengthen coordination between agencies on seafood safety and quality, particularly regarding imports. On July 29, 2009, the House approved H.R. 2749, a revised version of H.R. 759. H.R. 2749 takes a comprehensive approach to food safety, including seafood. On November 18, 2009, the Senate Health, Education, Labor, and Pensions Committee approved S. 510, another comprehensive approach to food safety.

Chapter 5- On June 28, 2007, FDA announced a broader import control of farm-raised catfish, basa, shrimp, dace (related to carp), and eel from China. FDA will start to detain these products at the border until the shipments are proven to be free of residues from drugs that are not approved in the United States for use in farm-raised aquatic animals. The agency took this action to protect American consumers from unsafe residues detected in these products. There have been no reports of illnesses to date.

FDA is taking this strong step now because of continuing evidence that certain Chinese aquaculture products imported into the United States contain illegal substances. Aquaculture, also known as fish farming, involves raising fish in enclosed areas to be sold for food. Almost half of all imported seafood is from aquaculture, according to the U.S. Department of Commerce

Chapter 6- FDA has statutory authority and responsibility for the safety of all food, except for most meats, poultry and processed egg products, which are regulated by the U.S. Department of Agriculture.

With respect to seafood, FDA operates a mandatory safety program for all fish and fishery products under the provisions of the Federal Food, Drug and Cosmetic (FD&C) Act, the Public Health Service Act, and related regulations. The FDA program includes research, inspection, compliance, enforcement, outreach, and the development of regulations and guidance. As a cornerstone of that program, FDA publishes the *Fish and Fisheries Products Hazards and*

Controls Guidance, an extensive compilation of the most up-to-date science and policy on the hazards that affect fish and fishery products and effective controls to prevent their occurrence. FDA is finalizing the fourth edition of this guidance document, which has become the foundation of fish and fishery product regulatory programs around the world.

Chapter 7- NOAA's National Marine Fisheries Service is responsible for the stewardship of the nation's living marine resources within the world's largest exclusive economic zone (EEZ). NOAA/NMFS protects and conserves these resources and their habitats through scientific research, fisheries management, efforts to protect marine mammals and endangered species, law enforcement, habitat conservation, and voluntary seafood quality and safety monitoring and inspection. NMFS has both domestic and international responsibilities, and seeks to maximize economic benefits from the sustainable use and conservation of living marine resources.

Chapter 8- Thank you for holding this hearing and inviting Food & Water Watch to testify on the food safety implications of imported Chinese seafood. China is a fisheries superpower and leads the world in total fish catch, fishing fleet size, aquaculture production and global seafood exports. China's fishing and aquaculture industry can have significant environmental impacts on the global health of ocean fisheries, China's environmental quality and the safety of seafood for consumers in China and here in the United States. China's vast aquaculture production poses special risks to food safety, since densely packed fish are often treated with antibiotics and veterinary medicines that are illegal to use on fish in America because of their associated human health and environmental risks.

In: Seafood Fraud and Safety: Background and... ISBN: 978-1-61761-500-9
Editors: Kaitlyn O. Harris

Chapter 1

SEAFOOD MARKETING: COMBATING FRAUD AND DECEPTION

Eugene H. Buck

SUMMARY

Congress is facing questions of whether the law applicable to fraudulent seafood sales and marketing is clear and enforceable, whether agency enforcement efforts targeting seafood fraud are adequate, and whether the penalties for seafood fraud are a deterrent. Congress may become involved in oversight of how federal agencies are addressing these issues, and legislation related to these concerns may be considered.

With increased seafood imports and decreased monitoring, fraud and deception in seafood marketing is becoming more widespread. The flesh of many fish species is similar in taste and texture and, therefore, it is difficult to identify species in fillet form, especially after preparation for consumption. Thus, it can be relatively easy to substitute an inexpensive species for one of higher value. Inaccurate (low) counts or net weights ("short weighting") result in consumers receiving less for their money than advertised and anticipated. Overbreading may cause consumers to pay shrimp prices for excess bread crumbs. Excessive amounts of glaze (overglazing) can deliberately be used to increase the apparent weight, and therefore the apparent value, of the delivered product. In addition, some new treatment procedures by the seafood industry, such as carbon monoxide/tasteless smoke, are being questioned for their potential to deceive consumers. Since food safety and bioterrorism concerns have taken precedence, regulatory agencies have been less able to maintain control of economic fraud.

The extent of this fraud is not well documented. The National Fisheries Institute (NFI) has undertaken an initiative to promote economic integrity within the seafood industry, concentrating on three primary areas:

- transshipment of products subject to antidumping and countervailing duties;
- mislabeling of products or species substitution; and
- mislabeling of weights or counts of products.

The Food and Drug Administration (FDA) is the primary agency responsible for ensuring that food sold in interstate commerce is properly labeled. FDA's jurisdiction covers seafood and the agency operates an oversight compliance program, the Seafood Regulatory Program, for fishery products. Responsibility for a food product's safety, wholesomeness, identity, and economic integrity rests with the processor or importer, who must comply with regulations promulgated under the Federal Food, Drug and Cosmetic Act (FFDCA) and the Fair Packaging and Labeling Act (FPLA).

In the 111th Congress, provisions were included in H.R. 759 and H.R. 2749 that proposed amending the FFDCA to restrict the use of carbon monoxide in meat, poultry, and seafood. In addition, S. 1406 was reported by the Senate Committee on Appropriations with report language encouraging FDA to more aggressively combat fraud in segments of the seafood industry.

This chapter reviews recent incidents of fraud and deception and examines related policy issues.

Media attention has focused on recent incidents of fraud relating to seafood—restaurants knowingly serving lower-priced fish than identified on menus, lower-priced species marketed commercially as higher-priced species, packaged weights of seafood less than labeled weights, and extra water added to seafood to increase total product weight—raising public concern. Most seafood fraud is based on supplying the consumer with something different from and inferior to the product expected.[1] In some instances, such practices may not be intentional, and the extent of fraud is not well documented. These occurrences have the potential to erode consumer confidence in seafood generally, potentially lowering Americans' consumption of seafood. Congress is facing questions of whether the law applicable to fraudulent seafood sales and marketing is clear and enforceable, whether agency enforcement efforts targeting seafood fraud are adequate, and whether the penalties for seafood fraud are a deterrent.

The primary federal law that addresses mislabeling is the Federal Food, Drug, and Cosmetic Act of 1938 (FFDCA; 21 U.S.C. §§ 301 et seq.), which is administered by the Food and Drug Administration (FDA). The FFDCA, as amended, gives FDA authority over most food regulation and includes (1) definitions elaborating on the concepts of adulteration and misbranding; (2) provisions for control over all labeling of foods in interstate commerce; (3) detailed regulation of safety and wholesomeness of foods; and (4) enforcement remedies available to the agency. In addition, the Fair Packaging and Labeling Act (FPLA; 15 U.S.C. § § 1451 et seq.) requires that consumers of packaged commodities be provided with accurate information as to its contents. Food products falling within the scope of the FFDCA that are introduced into interstate commerce in violation of the FPLA and its regulations are deemed to be misbranded within the meaning of the FFDCA. The Chemical Services Group of the National Marine Fisheries Service (National Oceanic and Atmospheric Administration, Department of Commerce) at the National Seafood Inspection Laboratory in Pascagoula, MS, provides analytical testing to identify fish species. Applicable law is discussed in detail below. Numerous issues concerning seafood safety also exist; for more analysis of this aspect, see CRS Report RS22797, *Seafood Safety: Background and Issues*, by Geoffrey S. Becker and Harold F. Upton.

FRAUDULENT OR DECEPTIVE PRACTICES

Mislabeling or Substituting Species

The differences in the taste and texture of different fish species' flesh are subtle, and therefore it is frequently difficult to identify a species in fillet form, especially after it is prepared for consumption. Thus, it is relatively easy to substitute an inexpensive species for one of higher value. Over the nine-year period of FY1 988-FY1 997,[2] routine examinations of seafood products by the National Marine Fisheries Service's National Seafood Inspection Laboratory (NSIL) found that 37% of fish and 13% of other seafood (e.g., shellfish, edible seaweed) from randomly selected vendors were mislabeled.[3]

Table 1. Examples of Commonly Substituted Seafood (less expensive products are in column B)

A	B
Red Snapper	Rockfish
Mahi Mahi	Yellowtail (*Serbola lalandb*)
Swordfish	Mako Shark
Orange Roughy	Oreo Dory or John Dory
Cod	Alaska Pollock
Halibut	Sea Bass
Dover Sole	Arrowtooth Flounder
Red Drum	Black Drum
Snapper (*Lutjanus* sp.)	Tilapia
Grouper	Basa or tra
Lake or Yellow Perch	White Perch or Zander
Caviar (Sturgeon species)	Paddlefish or other fish roe
Walleye	Sauger or Alaska Pollock
Chum Salmon	Pink Salmon
Salmon	Steelhead Trout
Pacific salmon	Atlantic salmon
Blue Crabmeat	Imported Crabmeat
Wild-Caught Salmon	Farm-Raised Salmon

Source: Table contents modified from http://www.cfsan.fda.gov/~frf/econ.html.

Identification of fish flesh usually requires DNA testing. There is debate within the seafood scientific laboratory testing community about the accuracy of public DNA sequence databases. Public databases seldom use validated fish standards and thus are likely to contain incomplete and/or inaccurate DNA sequences. Although most regulatory agencies are developing DNA methods for species identification,[4] some of these agencies continue to depend on a method that is unreliable for species identification of heat-processed or dried fish products—isoelectric focusing electrophoresis of soluble muscle proteins.[5] Some U.S. seafood importers and distributors employ private companies to conduct random testing to eliminate blatant species substitution and minimize unintentional species substitution. Although DNA methods for fish species identification are not yet officially recognized by FDA, the agency is working with an international group, FISHBOL,[6] to develop a qualified database of DNA sequences from taxonomically verified fish specimens. This database should be available for use within three years.

Fraud can be perpetuated in several ways. Unfair and deceptive trade practices occur when restaurants misrepresent menu items to their patrons by substituting other (often less desirable and less expensive) fish for an item described as a higher-valued species. Fraud also occurs at the manufacturing level, as in American Samoa, where six tuna cannery workers were convicted for their involvement in falsely labeling hundreds of cans as albacore tuna when the cans were actually filled with less expensive wahoo, and selling them to local stores.[7] Some distributors have knowingly sold restaurants and retailers lower-valued species, claiming that they are different species of higher value.

While species substitution may be intentional for certain species and products because of their differing values, it also occurs unintentionally where species identities are easily mistaken. Table 1 provides a list of some commonly substituted species. Additional instances of mislabeling include imported farmed salmon being falsely identified as wild Alaska salmon and frozen seafood being marketed as fresh product, as well as confusion over seafood being marketed as "organic." Since large quantities of seafood are imported for U.S. consumption, some portion of the mislabeling problem undoubtedly originates with foreign suppliers. To address this concern, FDA has issued an import alert on species substitution, providing guidance to agency field personnel regarding the manufacturers and/or products at issue.[8] Asian catfish (basa and tra) may illustrate the magnitude of the problem. Although import volumes exceed 50 million pounds annually for these species combined, they are difficult to find correctly identified in the domestic marketplace.[9]

Problems can arise from substitution at the restaurant level, misrepresentation by the restaurant supplier, or product misidentification anywhere in the harvesting and processing system. It is often difficult to determine who is at fault, especially if there is collusion. One recent study, based on an analysis of mitochondrial DNA sequences, revealed that 77% of the fish sold in the U.S. marketplace as red snapper, *Lutjanus campechanus*, belonged to other species of the Lutjanidae family, and could not legally be labeled as "red snapper."[10] In early 2007, Alabama agriculture inspectors targeted 35 Gulf of Mexico restaurants and seafood markets for sampling after a Mobile, AL, television station used DNA testing and found that only one in ten samples from restaurants advertised as grouper actually were this fish.[11] Alabama state law makes a restaurant or distributor subject to a fine of $5,000 per offense for selling falsely labeled seafood products. On September 3, 2008, the Florida Attorney General's Office announced a settlement with Sysco

Food Services—West Coast Florida, Inc., requiring the distributer to refrain from marketing to its restaurant clients any items described as grouper without taking commercially reasonable steps to determine that the item is genuine. [12] This legal action was initiated after a similar media investigative report in Tampa, FL, discovered that 17 of 24 area restaurants appeared to be selling less expensive species as grouper.[13] Media investigations have not been limited to the Gulf area.[14] Because the laboratories that tested fish samples for media inquiries all used DNA sequencing methods and made fish species identifications based on DNA sequences in public databases, there is debate within the seafood scientific laboratory testing community about the accuracy of these media reports. A more recent study of New York City restaurants and markets used DNA barcode technology[15] to determine that 14 of 56 samples collected were incorrectly labeled as higher- priced fish.[16]

Common Names of Fish

The correct use of names is crucial for properly identifying seafood. Because one species may have different vernacular names in different regions and because different species may be called by the same vernacular name, standard market names for seafood products are needed to avoid confusion. Common names of fish species have been standardized, often by professional societies.[17] FDA and the National Marine Fisheries Service (NMFS) have cooperated to develop "The Seafood List," compiling existing acceptable market names for imported and domestically available seafood. [18] In 1988, FDA published the FDA Guide to Acceptable Market Names for Food Fish Sold in Interstate Commerce ("The Fish List") to provide an authoritative source of common names to establish order in the marketplace and reduce confusion among consumers. In 1993, FDA published an updated, expanded "Seafood List," which includes invertebrate species (mollusks and crustaceans) as well as finfish. The frequently updated Seafood List reflects what FDA considers the most appropriate market names for the identification and labeling of seafood and is the agency's primary guidance for naming seafood sold in interstate commerce.

Two main provisions cover the naming of fish—Section 403 of the Federal Food, Drug, and Cosmetic Act of 1938 (FFDCA), which addresses misbranded food, and 21 C.F.R. § 102.5, which sets forth general principles for common or usual names for non-standardized food. A common or usual name is the "prevalent and meaningful name by which consumers ordinarily identify the food."[19] Use of either the acceptable market name or the common name in labeling seafood products assures that identity labeling of the seafood will comply with FDA and National Marine Fisheries Service regulations. FDA discourages the use of vernacular names as this practice may cause seafood to be misbranded.[20]

An example of this concern arose several years ago when increasing imports of basa (*Pangasius bocourti*) from Vietnam were marketed in the United States as "catfish," causing confusion with domestically produced *Ictalurid* catfish. FDA guidance formerly listed a number of fish other than those from the family *Ictaluridae* with the term "catfish" in their names. This prior guidance reflected what FDA believed were names for seafood that could be used by importers and domestic distributors and sellers consistent with the food naming provisions of the FFDCA. To address this confusion, § 10806 of P.L. 107-171 (Farm Security and Rural Investment Act of 2002) amended § 403 (the food misbranding provision) of the FFDCA (21 U.S.C. § 343) to provide that a food shall be deemed to be misbranded "[i]f it purports to be or is represented as catfish, unless it is fish classified within the family *Ictaluridae*." Although additional concern relates to the use of "lobster" in describing items generally identified as "langostinos,"[21] the extent of this problem is unknown.

Country-of-Origin Labeling

Mislabeling of the country of origin is another concern. Without labeling, consumers would rarely be able to distinguish a product's country of origin; labeling fish falsely as to country of origin removes the cachet from more desirable products, driving down the more desirable products' market price. Importers may falsely claim their seafood product is from a country where seafood products may be recognized as of higher quality than the actual source country. Particularly for seafood, some difficult questions relate to the meaning of the term "country of origin." For example, is product origin best represented by where the seafood was caught, the flag of the harvesting vessel, where the harvest was first landed, where the product was first processed, or where it was last or secondarily processed? Following the terrorist attacks on the World Trade Center on September 11, 2001, interest in traceability in the food supply chain increased substantially, including recognition of the advantages of a complete traceability program for seafood from producer to customer so that the source of and liability for any mislabeling can be more easily identified.[22]

In October 2004, the Agricultural Marketing Service (AMS), U.S. Department of Agriculture (USDA), promulgated an interim final rule requiring certain retailers and their suppliers to notify customers of the country of origin of wild and farm-raised fish and shellfish.[23] Final rules to implement fully the country-of-origin requirements were published by AMS on January 15, 2009.[24] USDA has entered into agreements with states having existing enforcement infrastructure to assist in compliance reviews for fish and shellfish. Only USDA is able to initiate enforcement actions against a person found to be in violation of the law, as this statute does not provide for a private right of action. NOAA is reportedly conducting a criminal investigation of alleged repacking and relabeling of imported farmed and wild shrimp as domestic U.S. product.[25] In April 2009, a Washington State man was fined $160,000 and sentenced to 30 days in jail for mislabeling 136,000 pounds of turbot from China as U.S. halibut.[26] For more information on country-of-origin labeling, see CRS Report RS22955, *Country-of-Origin Labeling for Foods*, by Remy Jurenas.

Low Weights or Undercounting

Inaccurate (low) counts or net weights ("short weighting") result in consumers receiving less for their money than advertised and anticipated. These instances, although commonly reported, also constitute mislabeling offenses under the FFDCA. The seafood community recognizes this problem to be of much greater concern than species substitution, costing legitimate businesses sales and reduced confidence in their true-packaged products. As part of its economic integrity initiative (see "Industry Initiatives," below), the National Fisheries Institute reported that, in the year preceding August 2008, it sent more than 80 letters to FDA documenting specific company examples of fraudulent weights.[27]

Over-Treating or Added Water Weight

The FDA standard for breaded shrimp requires that such a product contain at least 50% shrimp. Overbreading may cause consumers to pay shrimp prices for excess bread crumbs. Frozen fillets, shrimp, crab legs, and other products are normally protected from dehydration (freezer burn) while frozen by the application of a light glaze of ice, and the water weight of this glaze is not included in the stated product weight. Excessive amounts of glaze (overglazing), not compensated for in this manner, can deliberately be used to increase the apparent weight, and therefore the apparent value, of the delivered product. On February 20, 2009, FDA reissued its 1991 guidance, warning the seafood industry that short weight on a product obtained by including glaze in stated weights is a felony.[28]

Sodium tripolyphosphate (STPP) is used in the seafood industry to retain moisture in the product so as to prevent freezer burn and, when properly used, can be a legitimate means for aiding processing. However, STPP can be misused to retain excess moisture in seafood products. Prolonged soaking of seafood in an STPP-water solution can result, for example, in Atlantic sea scallops or shrimp with excessive water. Such excess water adds to the product's total weight, resulting in misrepresentation when seafood prices are charged for water and, in the case of shrimp, the product is bumped into a larger weight class where a higher price per pound can be charged. Seafood treated with STPP or other water-retaining chemicals must be accurately labeled to identify this treatment. However, FDA has found it difficult to set percent-moisture guidelines and labeling requirements for treated scallops,[29] and no moisture guidelines or standards exist for shrimp. Some critics suggest that the use of all water retention chemicals (primarily phosphates) should be prohibited in fish and shellfish.

Altered Color

Fish fillets can be treated with carbon monoxide (CO) to give fish flesh a fresher-appearing reddish tint.[30] The growing use of CO (also referred to as "tasteless smoke" or TS) as a "pigment fixative" has alarmed some consumer advocates who say it deceives shoppers who depend on color to help them avoid spoiled fish. Seafood industry representatives assert that TS/CO is a legitimate preservative, and the FDA has said it has no questions regarding an industry determination that TS/CO is a substance generally recognized as safe.[31] An additional consumer safety issue occurs when the flesh of certain species such as tuna develops toxic levels of histamine through time and/or temperature abuse—with CO treatment there are no visual cues to indicate when such flesh may be decomposed and toxic. Consumer advocates have urged FDA to conduct a formal evaluation of this treatment's impact on consumer safety. FDA considers tuna to be misbranded if it is treated with TS/CO but not labeled to indicate that it contains a preservative and thus purports to be unprocessed, fresh, or fresh-frozen tuna.[32] Thus, all processed seafood items involving TS/CO require label declarations under 21 C.F.R. Part

101.22(j).[33] In the 111th Congress, provisions have been introduced in H.R. 759 that would amend the FFDCA to restrict the use of CO in meat, poultry, and seafood.

Some aquaculture operations use the color additives canthaxanthin and/or astaxanthin in feed to impart a more orange color to fish flesh of salmon and/or trout.[34] The flesh of the farmed varieties of these fish would, if not for these color additives, be a less-appealing paler color. Under specified conditions, use of these additives in feed for salmonid fish is legal as long as fish are properly labeled to identify that this treatment has been used.[35] Farmed salmon and trout, where additives have been used to enhance color, are susceptible to being improperly labeled and this product deceptively and fraudulently marketed as “wild” fish.[36]

Transshipment to Avoid Import or Customs Duties

Transshipment occurs when foreign producers ship goods to a second country en route to the United States. Although transshipment is generally legal and commonly used in the ordinary course of business, it is illegal if it is done for the purpose of circumventing duties and other applicable trade restrictions. It has been reported that shrimp from China have been shipped to the United States by way of Indonesia to avoid paying antidumping duties of 112% levied by the United States on shrimp imported from China, but not on shrimp imported from Indonesia.[37]

Industry Initiatives

In 1986, the Southeastern Fisheries Association, funded under the Saltonstall-Kennedy Act grant program,[38] introduced a Seafood Product Quality Code program in one of the first efforts by a fisheries trade association to better inform wholesale and retail seafood buyers.[39] The code, an educational and promotional concept, catalogs the industry’s recommended measures for product quality for individual species to better inform buyers and to encourage more industry quality compliance by facilitating communications between buyers and sellers. Compliance with the code is a voluntary means whereby seafood sellers can indicate that they provide consumers a very safe, high-quality product.

In October 2006, concerns that seafood fraud had begun to and could increasingly erode consumer confidence in seafood led the National Fisheries Institute (NFI) to announce an initiative to promote economic integrity within the seafood industry. Implementation of this initiative commenced in summer 2007, concentrating on three primary areas:

- transshipment of products subject to antidumping and countervailing duties;
- mislabeling of products or species substitution; and
- mislabeling of weights or counts of products.

NFI officially launched its Better Seafood Bureau on July 5, 2007.[40] NFI is pursuing their initiative by (1) obtaining commitments from the CEOs of NFI member companies to comply with current law and regulation, and (2) developing an accountability system that would reward "good actors" and identify "bad actors." Such an accountability system involves screening by the Better Seafood Bureau, independent third-party audits of processes and products, and a member review process. The name was officially changed to the Better Seafood Board in May 2009 to avoid confusion with the Better Business Bureau.

CURRENT LAW AND CONGRESSIONAL INTEREST

With the increasing media attention to this issue, Congress may face questions concerning current law applicable to seafood marketing and fraud. The issues to consider may include whether:

- current law applicable to fraudulent seafood sales and marketing is clear and enforceable;
- federal agency enforcement efforts targeting seafood fraud are adequate;
- the penalties for seafood fraud offenses are a deterrent; and
- the resources for federal agency enforcement are sufficient.

FDA is the primary agency responsible for ensuring that food sold in interstate commerce is properly labeled. The agency's jurisdiction includes seafood and the agency operates an oversight compliance program, the

Seafood Regulatory Program, for fishery products. Responsibility for a food product's safety, wholesomeness, identity, and economic integrity rests with the processor or importer, who must comply with regulations promulgated under the FFDCA and the FPLA. The agency has the authority to detain or temporarily hold food being imported into the United States while it determines if the product is misbranded or adulterated. The agency also has the authority to take legal action against sellers of adulterated and misbranded seafood and to recommend criminal prosecution or injunction of responsible firms and individuals. However, enforcement of economic fraud and labeling laws may be a lower FDA priority relative to protecting the health and safety of the U.S. food supply.[41]

The adequacy of agency funding is also an issue, especially whether the agency has adequate resources so that it can more systematically monitor for economic fraud and mislabeling, better determine the scope and scale of these types of problems, and develop new programs that address these concerns. About 85 of the agency's roughly 1,350 inspectors work primarily with seafood. The adequacy of agency funding is regularly raised as a concern when agency appropriations are considered. For example, in testimony before the Senate Committee on Appropriations, one industry representative stated:

> We believe the most important issue on the table today is whether FDA is adequately funded, has sufficient staff with scientific training and experience ... , has research dollars available to address key questions, has strong working agreements with the States to support as needed and has the commitment of the President and the Congress.[42]

Federal Food, Drug, and Cosmetic Act

The FFDCA attempts to keep interstate commerce free from misbranded (i.e., mislabeled) articles and to protect the public from inferior foods resembling standard products but marketed under distinctive names.[43] The governing statute for naming food is the FFDCA. The FFDCA, as amended, gives FDA authority over most food regulation and includes:

- a series of definitions elaborating on the concepts of adulteration and misbranding;
- control over all labeling of foods traveling in interstate commerce;

- detailed regulation of issues concerned with safety and wholesomeness of foods; and
- enforcement remedies available to the agency, when needed.[44]

An article is deemed misbranded if, among other things, its labeling is false or misleading or it is offered for sale under the name of another food.[45] An article is considered mislabeled when the label makes "no representation as to definition and standard of quality," unless the label bears the common or usual name of the food, if there is one.[46]

FDA has issued regulations that outline general principles for common or usual names of food. The common or usual name must:

> accurately identify or describe, in as simple and direct terms as possible, the basic nature of the food or its characterizing properties or ingredients. The name shall be uniform among all identical or similar products and may not be confusingly similar to the name of any other food that is not reasonably encompassed within the same name. Each class or subclass of food shall be given its own common or usual name that states, in clear terms, what it is in a way that distinguishes it from different foods.[47]

A common or usual name of a food may be established by common usage or by regulation.[48] For example, FDA promulgated a regulation establishing that Pacific whiting or North Pacific whiting is the common or usual name of the food fish *Merluccius productus*.[49] Most common or usual names, however, are established through common usage.

Although the agency's Seafood List includes "vernacular" names for some species, use of vernacular names is discouraged by the agency. The use of a vernacular name may cause a seafood to be misbranded under the FFDCA.[50] For example, the agency has issued specific guidance on using "red snapper" as a market name. The agency's policy states that "the labeling or sale of any fish other than *Lutjanus campechanus* as red snapper constitutes a misbranding in violation of the [FFDCA]."[51]

Under the FFDCA, species substitution also violates FDA's prohibition against adulteration. A food is deemed adulterated "if any substance has been substituted wholly or in part."[52] The marketing of a less valuable fish as one of higher value is a substitution and can result in a finding of adulteration.

Fair Packaging and Labeling Act

The FPLA requires that consumers of packaged commodities be provided with accurate information as to its contents. Congress passed the FPLA to "enable consumers to obtain accurate information as to the quantity of the contents and to facilitate value comparisons."[53] Under the FPLA, it is unlawful for persons engaged in labeling or packaging of consumer commodities "to distribute or to cause to be distributed in commerce any such commodity if such commodity is contained in a package, or if there is affixed to that commodity a label, which does not conform to the provisions" of the act.[54]

The FPLA requires each label to identify the commodity and the name of its manufacturer, packer, or distributor and the net quantity of contents, in terms of weight or mass, measure, or numerical count.[55] Food products falling within the scope of the FFDCA that are introduced into interstate commerce in violation of the FPLA and its regulations are deemed to be misbranded within the meaning of the FFDCA.[56]

State Regulation of Seafood Labeling

In addition to federal requirements, some states also regulate the labeling and branding of seafood under state versions of the FFDCA. For example, in Alaska "no person may label or offer for sale any food fish product designated as halibut, with or without additional descriptive words, unless the food fish product is *Hippoglossus* or *Hippoglossus stenolepsis*."[57] In California, an individual who sells any commodity in less quantity than what is represented is guilty of a misdemeanor offense.[58]

Customs and Border Protection

Transshipment to avoid paying import or customs duties is illegal whenever it circumvents trade laws and other applicable trade restrictions.[59] The applicable law and regulation may vary, depending upon the trade agreement existing between the United States and another nation as well as the status of any antidumping and countervailing duties currently in force for particular products imported from designated nations.

CONCLUSIONS

Both consumers and some producers could incur economic losses in cases of fraud or deception, with short-term gains to those who sell mislabeled merchandise. Although it is not clear whether the amount of fraud and deception in seafood sales and marketing is increasing, media attention to this issue has raised its profile with the public. The economic integrity initiative of the National Fisheries Institute has the potential to increase attention within the seafood industry to this issue as well as to address eroding consumer confidence in fair marketing of seafood produce. In addition, third-party certification bodies—for example, the Marine Stewardship Council,[60] Naturland,[61] and the Aquaculture Certification Council[62]—have the potential to address fraud and deception concerns where certification requires producers to adhere to specific protocols and standards for catching/raising and processing product. However, the potential for these groups to self-enforce their standards has not been demonstrated.

In response to increased public concern, Congress is facing questions concerning current law applicable to seafood marketing and fraud. These questions include whether current law applicable to fraudulent seafood sales and marketing is clear and enforceable, whether federal agency enforcement efforts targeting seafood fraud are adequate, and whether the penalties for seafood fraud offenses are a deterrent. In addition, increased funding may be an issue so that agencies can more systematically monitor the situation, better determine the scope and scale of this type of problem, and develop new programs that address this fraud.

GAO released a report on seafood fraud in February 2009, which may provide impetus for additional attention to FDA's role in addressing this issue.[63] GAO concluded that federal agencies have not taken advantage of opportunities to share information that could benefit individual agency efforts to detect and prevent seafood fraud, and have not identified similar and sometimes overlapping activities that could be better coordinated to use limited resources more efficiently.

For additional background on broader legislative efforts to address seafood marketing and safety concerns, see the "Marketing" and "Seafood Safety and Nutrition" sections of CRS Report R40 172, *Fishery, Aquaculture, and Marine Mammal Issues in the 111th Congress*, by Eugene H. Buck and Harold F. Upton.

End Notes

[1] Ian Doré, *Seafood Scams and Frauds and How to Protect Yourself!* (Toms River, NJ: Urner Barry Publications, Inc., 1992).

[2] In September 2008, NSIL was contacted by CRS, but was unable to provide more recent statistics.

[3] See http://sst.ifas.ufl.edu/22ndAnn/file08.pdf.

[4] Rosalee S. Rasmussen and Michael T. Morrissey, "DNA-Based Methods for the Identification of Commercial Fish and Seafood Species," *Comprehensive Reviews in Food Science and Food Safety*, v. 7 (2008): 280-295.

[5] For a technical description, see http://www.chemsoc.org/ExemplarChem/entries/2003/leeds_chromatography chromatography/ief.htm; also see http://microbac.com/technical_articles/news_detail.php?news_ID=14.

[6] See http://www.fishbol.org/index.php. The Fish Barcode of Life (FISHBOL) Initiative is a global effort to assemble a standardized reference DNA sequence library for all fish species, and functions as a portal to the more comprehensive BoLD (Barcode of Life Database).

[7] "Wahoo Scam Yields Six Employees Sentenced to Probation," Samoa News (June 23, 2007); available at http://www.samoanews.com/monday/story3.html.

[8] For FDA import alert #16-04, see http://www.fda.gov/ora/fiars/ora_import_ia1604.html.

[9] For an example of mislabeling these imports, see http://agi.alabama.gov/press_releases/may-19-2006---sparks-warnsgulf-coast-businesses--restaurants-. See also *http://www.suffol kvanews.com/virginia/catfishconspiracy.php.*

[10] P.B. Marko, et al., "Mislabelling of a Depleted Reef Fish," *Nature*, v. 430(2004):309-310. FDA policy states that labeling or sale of any fish other than *Lutjanus campechanus* as red snapper constitutes a misbranding in violation of the FFDCA.

[11] See http://wkrg.com/news/article/whats_on_your_plate/3561/ and http://wkrg.com/news/article/is_it_really_grouper/ 3557/.

[12] See *http://myfloridalegal.com/newsrel.nsf/newsreleases/ C67E4A15E6C1A3EA852574B9004 CF240.*

[13] See http://pqasb.pqarchiver.com/sptimes/access/1230334761.html?dids= 1230334761: 1230334761&FMT=FT& FMTS=ABS:FT&date=Mar+10%2C+2007&author=CURTIS+ KRUEGER&pub=St.+Petersburg+Times&edition=&startpage=1 .A&desc=%27Grouper%27+costs+restaurants.

[14] Others have been reported in Portland, OR, http://www.kptv.com/print/11072970/detail.html; and in Phoenix, AZ, http://www.kpho.com/print/10371007/detail.html, for example.

[15] DNA barcoding is a relatively new technique that uses a short DNA sequence from a standardized and agreed-upon position in the genome as a molecular diagnostic for species-level identification. For additional information, see http://www.barcoding.si.edu/.

[16] John Schwartz, "Fish Tale Has DNA Hook: Students Find Bad Labels," *The New York Times*, August 22, 2008. Available at http://www.nytimes.com/2008/08/22/science/22fish.html.

[17] For example, see American Fisheries Society, *Common and Scientific Names of Fishes from the United States, Canada, and Mexico*, Special Publication 29, 6th edition (Bethesda, MD: July 2004), 386 p.; and American Fisheries Society, *World Fishes Important to North Americans*, Special Publication 21 (Bethesda, MD: 1991), 243 p.

[18] See http://www.cfsan.fda.gov/~frf/seaintro.html.

[19] 63 *Fed. Reg.* 20,148 (April 23, 1998).

[20] Updated information on FDA concerns related to species substitutions and misbranding can be found at http://www.fda.gov/ora/fiars/ora_import_ia1604.html.

[21] See http://www.lawfuel.com/show-release.asp?ID=3593.

[22] Ami Petersen and David Green, *Seafood Traceability: A Practical Guide for the U.S. Industry*, North Carolina Sea Grant, Report UNC-SG-06-04, 28 p.; available at http://www.ncseagrant.org/files/seafood

[23] 7 *C.F.R.* Part 60; See 69 *Fed. Reg.* 59708-59750 (October 5, 2004).

[24] 74 *Fed. Reg.* 2658-2707.

[25] See http://www.nola.com/news/?/base/news-2/1246253416178870.xml&coll=1.

[26] See http://www.noaanews.noaa.gov/stories2009/20090511_seafood

[27] John Sackton, Seafood.Com *News*, August 12, 2008.

[28] See http://www.cfsan.fda.gov/~dms/seaglgui.html.

[29] See http://www.fda.gov/bbs/topics/ANSWERS/ANS00422.html.

[30] See http://www.sushiman.net/ahi/carbon _monoxide _trea.htm.

[31] See CRS Report RL34247, *Federal Regulation of Substances Generally Recognized As Safe (GRAS) and the Use of Carbon Monoxide in Packaging for Meat and Fish*, by Vanessa K. Burrows and Cynthia Brougher.

[32] See FDA Import Bulletin #16B-95, available at http://seafood

[33] See http://www.cfsan.fda.gov/~rdb/opa-g015.html.

[34] Fereidoon Shahidi, Metusalach, and Joseph A. Brown, "Carotenoid Pigments in Seafoods and Aquaculture," *Critical Reviews in Food Science and Nutrition*, v. 28, no. 1 (1998): 1-67.

[35] 21 C.F.R. § 73.35; 21 C.F.R. § 73.75.

[36] A 2005-2006 study found that only 10 of 23 salmon fillets marketed as "wild" salmon were definitely caught in the wild, while the majority were from farmed salmon. See "Mislabeled Salmon: The Salmon Scam," *Consumer Reports*, v. 71, no. 8 (August 2006): 15.

[37] See http://www.atimes.com/atimes/Southeast_Asia and *http://au.sys-con.com/node/890860/print*. Further information is contained in out-of-print CRS Report RS2 1776, *Shrimp Trade Dispute: Chronology*, by Eugene H. Buck, available from the author.

[38] For further information, see CRS CRS Report RS2 1799, *Saltonstall-Kennedy Fishery Funding*, by Eugene H. Buck.

[39] See http://www.southeasternfish.org/SPQC/product%20quality%20code.pdf.

[40] For more information, see http://www.aboutseafood.com/about/about-nfi/better-seafood and http://www.aboutseafood.com/sites/www.aboutseafood.com/files/share/BSB_Backgrounder_General_May_2009_NRA.pdf.

[41] "Species Substitution: Labeling Law Not An FDA Priority ... ," *Santa Monica Seafood SeaLog* (April 2006).

[42] Testimony by Thomas E. Stenzel, President and CEO of the United Fresh Produce Association, before the Senate Committee on Appropriations, March 12, 2007, in a special hearing on "Food Safety.".

[43] 35A Am. Jur. 2d Food § 25 (2006).

[44] For more information, see CRS Report RL33559, *Food Safety: National Uniformity for Food Act*, by Donna V. Porter.

[45] 21 U.S.C. § 343. FDA further regulates this area of "Misbranding" in 21 C.F.R. § 101.18.

[46] 21 U.S.C. § 343(i)(1).

[47] 21 C.F.R. § 102.5(a).

[48] 21 C.F.R. § 102.5(d).

[49] 21 C.F.R. § 102.46.

[50] Sec. 540.750 Common or Usual Names for Seafood in Interstate Commerce (CPG 7108.26).

[51] Sec. 540.475 Snapper—Labeling (CPG 7108.21).

[52] 21 U.S.C. § 342 (b)(2).

[53] 15 U.S.C. § 1451.

[54] 15 U.S.C. § 1452.

[55] 15 U.S.C. § 1453(a).

[56] 15 U.S.C. § 1456(a).

[57] Alaska Stat. § 17.20.045.

[58] Cal Bus & Prof Code § 12024.

[59] Section 592 of the Tariff Act of 1930 (19 U.S.C. § 1592).
[60] See http://www.msc.org/about-us.
[61] See http://www.naturland.de/certification
[62] See http://www.aquaculturecertification.org/.
[63] U.S. Government Accountability Office, *Seafood Fraud: FDA Program Changes and Better Collaboration among Key Federal Agencies Could Improve Detection and Prevention*, February 2009, GAO-09-258, 49 p.

In: Seafood Fraud and Safety: Background and… ISBN: 978-1-61761-500-9
Editors: Kaitlyn O. Harris

Chapter 2

Seafood Fraud: FDA Program Changes and Better Collaboration among Key Federal Agencies Could Improve Detection and Prevention

United States Government Accountability Office

Why GAO Did This Study

In 2007, Americans consumed almost 5 billion pounds of seafood. Most seafood buyers, at many levels—importers, distributors, supermarkets, restaurants, and individual consumers—assume that the seafood they buy is what the seller claims it is. However, this is not always the case. Sometimes seafood products are mislabeled for financial gain—an activity called seafood fraud. Three federal agencies play key roles in detecting and preventing seafood fraud: the Department of Homeland Security's Customs and Border Protection (CBP), the Department of Commerce's National Marine Fisheries Service (NMFS), and the Department of Health and Human Services' Food and Drug Administration (FDA). GAO was asked to determine (1) the actions key federal agencies take to help detect and prevent seafood fraud and (2) the extent to which these key federal agencies collaborate with each other to help detect and prevent seafood fraud. GAO reviewed data and documents from each agency on actions to detect and prevent seafood fraud, and interviewed agency officials and other key stakeholders.

WHAT GAO RECOMMENDS

GAO is making recommendations to CBP, NMFS, and FDA that are intended to help reduce the prevalence of seafood fraud, increase interagency collaboration, improve information sharing, and reduce overlaps. In commenting on a draft of this chapter, CBP, NMFS, and FDA generally agreed with the recommendations.

WHAT GAO FOUND

CBP and NMFS conduct several activities to help detect and prevent seafood fraud, but FDA told GAO that it focuses on food safety and undertakes few fraud-related activities. Nonetheless, fraud can result in food safety problems. For example, fish that was mislabeled as a different species for financial gain has caused illnesses due to the presence of a potentially deadly toxin.

- CBP reviews seafood import documentation to detect schemes to avoid paying the appropriate customs duties as seafood products enter the country, among other things.
- NMFS addresses seafood fraud through its voluntary, fee-for-service inspection program, which includes inspecting seafood that retailers, among others, are purchasing to verify its net weight and ensure the species is correctly identified. According to NMFS officials, NMFS inspects approximately one-third of the seafood consumed in the United States.
- FDA examines only about 2 percent of imported seafood annually, and its primary seafood oversight program does not address economic fraud risks, which limits its ability to detect fraud. An FDA seafood fraud-related activity is the maintenance of a publicly available list of seafood names that is intended to help the industry correctly label products. However, until 2009, FDA had not fully updated the list it created in 1993 to reflect over 400 name changes. Finally, FDA's

guidance to help seafood processors comply with its seafood oversight program does not reflect the seafood labeling requirement of the Food Allergen Labeling and Consumer Protection Act of 2004 to include the species of fish or shellfish on product labels. Because of the limited scope of FDA's seafood oversight program, its mismanagement of the Seafood List, and its failure to update its guidance to reflect the allergen labeling requirement, consumers have less assurance that the seafood they purchase is correctly labeled.

The federal agencies that share responsibility for detecting and preventing seafood fraud—CBP, NMFS, and FDA—do not effectively collaborate with each other. Specifically, they have not identified a common goal, established joint strategies, or agreed on roles and responsibilities. As a result, the agencies have not taken advantage of opportunities to share information that could benefit each agency's efforts to detect and prevent seafood fraud, nor have they identified similar and sometimes overlapping activities that could be better coordinated to use limited resources more efficiently. For example, each agency has its own laboratory capability for determining seafood species and uses different methodologies for creating standards for species identification. The result is that neither the laboratories nor the data developed in them are shared.

ABBREVIATIONS

CBP	Customs and Border Protection
FDA	Food and Drug Administration
HACCP	Hazard Analysis and Critical Control Point
HHS	Department of Health and Human Services
ICE	Immigration and Customs Enforcement
MOU	memorandum of understanding
NFI	National Fisheries Institute
NMFS	National Marine Fisheries Service
NOAA	National Oceanic and Atmospheric Administration
NTAG	National Targeting and Analysis Group
USDA	U.S. Department of Agriculture

February 19, 2009

The Honorable Olympia J. Snowe
Ranking Member
Subcommittee on Oceans, Atmosphere, Fisheries, and Coast Guard
Committee on Commerce, Science, and Transportation
United States Senate

Dear Senator Snowe:

In 2007, Americans consumed almost 5 billion pounds of seafood. Most seafood buyers, at many levels—importers, distributors, supermarkets, restaurants, and individual consumers—assume that the seafood they buy is what the seller claims it is. However, this is not always the case. Sometimes people mislabel seafood products for financial gain—an activity called seafood fraud. The most common types of seafood fraud are shipping products through an intermediary country to avoid customs duties (transshipping), adding excessive amounts of water or ice to the seafood to increase its weight (over-treating), substituting a different species of seafood for the species listed on the label (species substitution), including less seafood in a package than indicated by the label (short- weighting), and other types of mislabeling or misrepresenting of seafood products.

Although comprehensive information on the extent of seafood fraud does not exist, seafood industry officials believe that seafood fraud is a problem. The available information suggests that the scope and economic impact of seafood fraud vary widely and can occur at any point in the seafood supply chain, from large-scale, multinational schemes involving importers—with impacts in the millions of dollars—to fraudulent activities at individual restaurants or grocery stores. Seafood fraud can have both food safety and economic consequences. For example, seafood fraud affected food safety in 2007 when imported puffer fish, which contains a potentially deadly toxin, was mislabeled as monkfish, and people became ill. Another example of fraud is including the ice-glaze covering used to freeze and preserve fish fillets in the net weight of the product, which has primarily an economic impact because consumers get less fish than they paid for. Seafood fraud can undermine consumer confidence in the U.S. seafood supply, over 80 percent of which is imported.

Three federal agencies play key roles in detecting and preventing seafood fraud: the Department of Homeland Security's Customs and Border Protection (CBP), the Department of Commerce's National Marine Fisheries Service (NMFS), and the Department of Health and Human Services' Food and Drug

Administration (FDA). CBP collects customs duties on imports, including seafood, and seeks to prevent the evasion of customs duties. Goods imported into the United States may be subject to duties on the basis of their product type, value, and country of origin, among other things. In addition, to limit the sale of foreign-made products in this country at less-than-normal value, called "dumping," some imported goods are also subject to antidumping duties. NMFS provides fee-forservice inspection services on request to the seafood industry, including processors, distributors, and other firms. These inspections can address economic integrity issues, such as the accuracy of a seafood product's label, as well as seafood safety issues. Finally, FDA is responsible for ensuring that the nation's food supply, including seafood, is safe, wholesome, and properly labeled. To that end, FDA is authorized to issue regulations to enforce the Federal Food, Drug, and Cosmetic Act. The act prohibits the misbranding or adulteration of food products, which would include the mislabeling and substituting of seafood products that constitute seafood fraud. FDA is responsible for seafood that is imported into the United States as well as seafood that is harvested and processed domestically. FDA inspects U.S. importers and domestic and foreign processors to ensure their compliance with applicable requirements, including labeling requirements and FDA's Seafood Hazard Analysis and Critical Control Point (HACCP) regulations. The HACCP regulations require seafood processors to identify and develop processes to mitigate biological, chemical, and physical hazards that are likely to occur. FDA also provides guidance to the seafood industry on the naming of seafood products and on food safety hazards. Table 1 shows the types of seafood fraud that these three agencies might detect when performing their authorized roles.

Table 1. Types of Seafood Fraud That Federal Agencies Might Detect

Fraud type	CBP	NMFS	FDA
Transshipment to avoid duties	√		
Over-treating		√	√
Species substitution	√	√	√
Short-weighting		√	√
Other mislabeling or misrepresenting	√	√	√

Source: GAO analysis of documents obtained from and discussions with CBP, NMFS, and FDA.

When an issue crosses federal agency lines, as seafood fraud does, the agencies involved must collaborate to deliver results more efficiently and effectively. For the purpose of this chapter, we define "collaboration" as a joint effort by two or more agencies that is intended to produce a greater public benefit than when the agencies act alone. Our previous work indicated that federal agencies can efficiently and effectively collaborate when they, among other things, (1) identify a common goal, (2) establish joint strategies to achieve that goal, (3) agree on their roles and responsibilities, (4) identify ways to maximize and leverage their resources, and (5) establish procedures and policies for working together systematically across agency lines.[1] When agencies do not collaborate efficiently and effectively, their individual efforts are carried out in an uncoordinated way, thereby limiting the overall effectiveness and efficiency of federal expenditures. For example, in 2007, we added the federal oversight of food safety to our high-risk list because this fragmented system—in which 15 federal agencies collectively administer at least 30 laws related to food safety—has caused inconsistent oversight, ineffective coordination, and inefficient use of resources.[2]

In this context, you asked us to determine (1) the actions key federal agencies take to help detect and prevent seafood fraud and (2) the extent to which these key federal agencies collaborate with each other to help detect and prevent seafood fraud.

Our research identified CBP, NMFS, and FDA as the key agencies involved in detecting and preventing seafood fraud. To determine the actions these agencies have taken to detect and prevent seafood fraud, we reviewed data and documents from each agency on the amount and nature of seafood fraud that they have identified, actions they have taken to prevent seafood fraud, and actions they have taken against fraud perpetrators. At all three agencies, we reviewed program guidance, inspection operation manuals, and other documentation and interviewed knowledgeable officials. We also obtained examples of seafood fraud criminal investigation cases and met with representatives from the major seafood industry trade associations. To determine the extent to which key federal agencies collaborate with each other to help detect and prevent seafood fraud, we reviewed existing federal interagency agreements and spoke with agency officials from CBP, NMFS, and FDA. (App. I provides additional information on our scope and methodology.)

We conducted this performance audit from January 2008 to February 2009, in accordance with generally accepted government auditing standards. Those standards require that we plan and perform the audit to obtain sufficient, appropriate evidence to provide a reasonable basis for our findings and conclusions based on our audit objectives. We believe that the evidence obtained provides a reasonable basis for our findings and conclusions based on our audit objectives.

RESULTS IN BRIEF

CBP and NMFS conduct several activities to help detect and prevent seafood fraud; however, FDA has taken few actions because it sees food safety as its top priority. These agency activities are described in the following text:

- CBP focuses on detecting schemes to avoid paying customs duties as seafood products enter the country, such as transshipment to avoid antidumping duties. CBP's import specialists review seafood import documentation on product type, value, and country of origin to ensure that importers have paid the appropriate duties. The agency also uses information provided by one of its National Targeting and Analysis Groups to help identify potentially fraudulent seafood shipments. This group analyzes data on foreign producers and importers that may be involved in transshipment schemes to avoid paying antidumping duties and works with port officials to examine these shipments as they arrive. For example, Chinese shrimp have been subject to an antidumping duty since 2005 because producers have set prices on the shrimp that were lower than the normal value. On the basis of this information and allegations from the domestic shrimp industry, a National Targeting and Analysis Group began to scrutinize imports of shrimp from Chinese producers and identified approximately $6 million worth of Chinese shrimp that had been transshipped through Indonesia in 2005 to avoid antidumping duties.
- NMFS's seafood fraud detection activity consists of a voluntary, fee-forservice inspection program that supports the seafood industry in two ways. First, seafood retailers, such as supermarkets, may ask NMFS to inspect the seafood products they purchase to ensure the products have not been misrepresented. In such cases, NMFS's inspectors verify a product's net weight and ensure the species is correctly identified. Second, seafood processors may ask NMFS, through its Quality Management Program, to inspect their seafood

processing operations to not only ensure compliance with FDA's HACCP regulations but also to identify measures that can help prevent seafood fraud. Such measures may include requiring the processor to weigh all products and compare that information with the stated weight on the package before the product leaves the facility and periodically testing the scales. According to officials in NMFS's Seafood Inspection Program, NMFS inspects approximately one-third of the seafood consumed in the United States.

- FDA sees ensuring the safety of the nation's food supply as a top priority and, therefore, devotes minimal resources to detect and prevent seafood fraud. Nevertheless in the course of conducting its food safety activities, FDA has at times incidentally uncovered seafood fraud. More specifically, as part of its food safety activities to ensure that imports are not contaminated, among other things, FDA examines imported seafood products, which occasionally has resulted in its identifying seafood fraud. However, FDA's opportunities to identify fraud are limited because it examines only about 2 percent of imported seafood. FDA is also limited in its ability to detect seafood fraud because its primary oversight program for seafood processors—HACCP—does not require them to identify and mitigate economic fraud risks that can occur during processing. However, seafood oversight programs operated by NMFS and the Canadian government include such requirements. An FDA seafood fraud-related activity is the maintenance of a publicly available list of scientific and market names of seafood—the Seafood List—that is intended to help the seafood industry comply with FDA's regulations on product mislabeling. However, FDA did not fully update the publicly available list it created in 1993 until January 2009, and does not provide stakeholders with an opportunity to comment on proposed changes before they are finalized. In addition, FDA provides guidance to seafood processors to help ensure that their seafood products are safe. However, this guidance does not reflect the seafood labeling requirements in the Food Allergen Labeling and Consumer Protection Act of 2004, which requires that the species of fish or shellfish be included on product labels to notify consumers who may be allergic to a particular species of fish. According to a senior FDA official, the act's labeling requirements also could help detect and prevent species substitution, since processors would need to verify the species of fish or shellfish to ensure accurate labeling. Because the seafood HACCP

> regulations do not address economic fraud risks; the Seafood List, until very recently, had not been kept up to date; and FDA's guidance does not reflect the allergen labeling requirements, the seafood industry may be less vigilant in ensuring their seafood products are correctly labeled and the public may more often encounter seafood products that are not what they are advertised to be. Consequently, we are recommending that FDA improve its ability to detect and prevent seafood fraud by (1) proposing amendments to its HACCP regulations to include measures to identify and mitigate economic fraud risks; (2) providing the opportunity for stakeholder comments prior to formalizing any changes to the Seafood List and routinely updating the public list; and (3) updating its guidance to reflect the seafood labeling requirements of the Food Allergen Labeling and Consumer Protection Act of 2004.

The federal agencies that share responsibility for detecting and preventing seafood fraud—CBP, NMFS, and FDA—do not efficiently and effectively collaborate with each other, which can diminish the efficiency and effectiveness of their efforts. Specifically, these agencies have not worked together to identify a common goal related to seafood fraud, established joint strategies to achieve such a goal, or agreed on their roles and responsibilities. Moreover, they have not identified ways to maximize and leverage their resources or established processes and policies for working together systematically across agency lines. As a result, these agencies are not sharing important information that could be helpful in detecting and preventing seafood fraud. For example, when FDA reviews the labels of imported products to identify potentially fraudulent labeling, it does not systematically share the results of these reviews with CBP. If CBP had access to the results of the labeling reviews, it could compare this information with the labels on products entering the country and better determine whether a product was mislabeled to avoid a customs duty or other import restrictions. In addition, these agencies have not leveraged their resources to address seafood fraud efficiently and effectively. For example, NMFS's voluntary fee-for-service inspection program and FDA's health and safety inspections are similar. However, an FDA official said that the agency is not sure whether it can rely on NMFS inspections, in part due to concerns about potential conflicts of interest, because NMFS is paid by industry to conduct its inspections. FDA raised this same concern to us in 2004 but added that it already had agreements with NMFS to deal with seafood safety and inspections, and that it would look at other ways to better leverage NMFS resources. Nonetheless, FDA does not currently try to determine whether NMFS has already inspected a seafood facility when it is deciding which facilities to inspect. Consequently, some facilities may be "over-inspected," while others are not inspected frequently enough. For example, in fiscal year 2007, FDA inspected 104 seafood facilities that were also inspected by NMFS; while FDA had not inspected 1,464 other facilities since before fiscal year 2003. In addition, CBP, NMFS, and FDA each has its own laboratory capability for, among other things, determining the species of seafood samples they receive. The agencies also use different testing methodologies and standards for species identification and do not acknowledge each other's laboratory results, nor do they generally share the species standards they have developed. As a result, resources are not used efficiently or effectively. We are recommending that CBP, NMFS, and FDA collaborate to (1) develop goals, strategies, and mechanisms to efficiently and effectively share information and resources related to seafood fraud detection and prevention across agency boundaries and (2) create a federal agencywide library of seafood species standards. In commenting on a draft of this chapter, the Department of Commerce, representing NOAA; the Department of Health and Human Services, representing FDA; and the Department of Homeland Security, representing CBP, generally agreed with our recommendations. Appendixes II, III, and IV contain reprints of the departments' letters, respectively.

BACKGROUND

Seafood fraud can include a variety of illegal activities done for economic gain and can occur at any point in the seafood supply chain. The domestic seafood supply chain begins with the harvester—that is, the people who catch or farm the seafood. From there, seafood products are shipped to processors, which then produce fresh, frozen, breaded, or cooked seafood. Processors or distributors then sell the seafood to supermarkets or restaurants. The process is similar for foreign seafood products, with one exception: these products enter the country through an importer and then move on to a distributor. According to the National Oceanic and Atmospheric Administration (NOAA), in fiscal year 2007, over 80 percent of the seafood consumed in this country was imported, and shrimp was the most widely consumed seafood. Seafood imports into the United States most frequently come from Canada, China, and

Thailand, according to the U.S. Department of Agriculture's (USDA) Foreign Agriculture Service.

Federal investigations have identified incidents of seafood fraud. For example, the Department of Homeland Security's Immigration and Customs Enforcement (ICE)[3] and NOAA's Office for Law Enforcement investigation that began in 2004 identified seafood fraud that involved smuggling and distributing mislabeled catfish into the United States from Vietnam. According to the indictment, an individual and his companies in Florida aided by exporters in Vietnam imported thousands of pounds of catfish into the United States labeled as grouper to avoid paying antidumping duties that the Department of Commerce had imposed on Vietnamese catfish. The defendant pled guilty and was sentenced to 51 months in prison. The judge also ordered the companies to forfeit property and pay over $1 million in restitution.

Seafood fraud can include a variety of illegal activities undertaken for economic gain. Such fraud typically involves mislabeling the seafood product and can include the following actions:

- *Transshipment to avoid duties*: Foreign producers may ship seafood products on route to the United States through a third country to avoid import duties by labeling the product's country of origin as the third country and also to avoid regulatory controls such as FDA import alerts.[4]
- *Over-treating*: Processors may, for example, over-bread prepared seafood products, use water-retaining chemicals, or over-glaze with an ice covering to artificially increase the weight of seafood products without indicating the true net weight of the seafood on the label.
- *Species substitution*: Participants in the seafood supply chain may label a species of seafood as another species. Typically, a lower-market-value species is labeled as a higher-market-value species to realize a larger profit. This results in consumers paying too much for the product.
- *Short-weighting*: Participants in the seafood supply chain may label packages of seafood as containing more than they actually contain.
- *Other mislabeling and misrepresenting*: Participants in the seafood supp If you want truly to understand something, try to change it *http://rnd009.googlepages.com/google* seafood product or can commingle two or more different products having different values but sell the entire lot at the value of the highest priced product.

Seafood fraud is an inherently deceptive activity and poses challenges for federal agencies to detect and prevent it while still maintaining the flow of legitimate seafood goods into and within the United States. Some fraudulent activity can even occur openly. For example, seafood companies receive public, written solicitations to purchase fraudulent seafood products. One type of solicitation offers to sell packages of fish fillets that are purposely mislabeled as another fish type to avoid antidumping duties and also capitalize on the higher market value of the falsely labeled fish type. These fish fillets are sold at a discount to the initial buyer, but then can be fraudulently resold for a higher price. Another type of solicitation offers to sell short-weight packages of seafood at a discount that the buyer could then fraudulently resell at the full price on the basis of the labeled package weight.

Federal, state, and local agencies play a role in detecting and preventing seafood fraud throughout the supply chain. In general, federal agencies inspect seafood processors, distributors, and importers and imported seafood products. States also inspect seafood processors either through contracts with FDA or under their own authority. States and local governments inspect and regulate retail establishments such as restaurants and supermarkets.

CBP's trade-related responsibilities include assessing the final customs duties, including antidumping duties, due on imports and collecting those duties. All goods imported into the United States are subject to a rate of duty, which may be free for certain products. The U.S. government has established a duty rate according to the product classification code—for example, fish sticks and other fillets that are breaded or coated with batter have a specific classification code and duty rate.[5] Antidumping and countervailing duties may also be required on imported products. The U.S. government may impose antidumping duties on products exported to the United States at unfairly low prices (i.e., dumping) and countervailing duties on products exported to the United States that were subsidized by foreign governments. The Department of Commerce sets these duty rates for specific products, countries or manufacturers, or any combination of these factors.

NMFS provides voluntary fee-for-service inspection services through its Seafood Inspection Program on request to the seafood industry, primarily under the authority of the Federal Agricultural Marketing Act of 1946. NMFS provides its services to domestic and foreign seafood firms to help them ensure compliance with all applicable federal food regulations, including FDA's seafood HACCP regulations. NMFS's services include inspections of a firm's processes and products to identify food safety and economic fraud risks as well as laboratory analyses.

Along with its responsibility for ensuring the safety of other food products under the Federal Food, Drug, and Cosmetic Act, FDA has the primary responsibility for ensuring that the nation's seafood is safe, wholesome, and properly labeled.[6] To carry out these responsibilities, FDA has created oversight programs designed to, among other things, examine and sample imported seafood products, inspect domestic and foreign seafood processors and importers, and assist state and local governments in their efforts to regulate retail establishments such as restaurants and supermarkets. Under the HACCP regulations, seafood processors are required to prepare and maintain a plan identifying critical points in the processing where contamination is reasonably likely to occur and implement control techniques to prevent or mitigate the contamination. FDA then inspects U.S. importers and domestic and foreign seafood processors to ensure their compliance with these HACCP regulations. When FDA first proposed the seafood HACCP regulations in 1994, the agency recommended that HACCP plans include controls for nonsafety hazards such as economic adulteration. Specifically, FDA recommended that the seafood industry adopt preventive processing measures to help ensure that, among other things, seafood was correctly identified and its weight properly recorded. Furthermore, FDA also proposed guidelines on how a seafood processor could use a HACCP-based approach to ensure that fish and fishery products were in compliance with the economic adulteration and misbranding provisions of the Federal Food, Drug, and Cosmetic Act. However, in finalizing its HACCP regulations, FDA eliminated these economic fraud controls. In discussing its reasoning for eliminating the economic fraud controls in the *Federal Register*, FDA stated that the seafood HACCP system would need to mature before the agency could determine whether it should address matters other than food safety hazards, such as economic fraud.

The seafood industry also plays a role in detecting and preventing seafood fraud. Seafood industry associations represent various aspects of the industry throughout the supply chain, from harvesters to retailers. They include product-specific associations, such as the Southern Shrimp Alliance in southern states or the Maine Lobster Promotion Council, or business-specific associations, such as the National Restaurant Association. These industry associations work to protect the brand name and quality of their respective products or businesses; monitor issues and legislation that may impact the industry, including fraud and other illegal activities; and work with government agencies and other organizations to promote the health and viability of their industry. Some of these associations have also provided federal agencies with information on potential seafood fraud, such as transshipping schemes to avoid antidumping duties and advertisements for short-weighted seafood products. In addition, the National Fisheries Institute (NFI) also created the Better Seafood Bureau to help its members combat such seafood fraud problems as transshipping to avoid antidumping duties, species substitution, and short-weighting.[7] According to NFI, the goal of the Better Seafood Bureau is to promote economic integrity in the seafood industry and assure customers of the quality of the institute's members' products. NFI requires a hand-signed contract from each member's Chief Executive Officer that he or she will comply with all U.S. laws and regulations, and has created an accountability system that requires members who break the contract to pay for a third-party audit of its processes.[8] The Southeastern Fisheries Association also developed the Seafood Product Quality Code in 1984 to educate any interested harvesters, processors, distributors, wholesalers, retailers, and consumers on standards and measures to promote product quality and build confidence in the seafood industry.

Federal agencies face challenges in achieving their missions, especially if they have limited resources and some elements of their missions are shared with other agencies. Effective collaboration is often key to overcoming these challenges, and our previous work has identified practices that can help enhance and sustain collaboration.[9] First, collaboration requires agency staff, working across agency lines, to define and articulate the common federal outcome or goal they are trying to achieve. Second, once a common goal is established, agencies need to develop joint strategies. Such strategies help align the partner agencies' individual activities and resources to contribute to accomplishing the common goal. Third, the agencies should work together to agree on their roles and responsibilities. By agreeing on their roles and responsibilities, agencies can clarify who will do what, organize their joint and individual efforts, and determine who will lead the collaborative effort. Fourth, because each agency may contribute different activities and levels of resources toward achieving the common goal, by assessing their relative strengths and weaknesses, collaborating agencies can identify opportunities to leverage each other's resources. This may lead to additional benefits that would not have been available if they were working separately. Fifth, to ensure consistent implementation of their activities and a sustained collaborative effort to achieve the common goal, the agencies should develop compatible policies and procedures for all of the agencies to follow. (App. I includes a list of the three other collaboration practices we previously identified but did not address in this chapter.)

CBP AND NMFS TAKE A VARIETY OF ACTIONS TO DETECT AND PREVENT SEAFOOD FRAUD, WHILE FDA TAKES FEW ACTIONS RELATED TO SEAFOOD FRAUD

CBP focuses on detecting schemes to avoid paying customs duties by reviewing import information; targeting and, along with ICE, investigating potential seafood fraud perpetrators; and taking enforcement actions, if warranted. NMFS offers a voluntary, fee-for-service seafood inspection program that can detect seafood fraud, such as short-weighting, and may also help prevent seafood fraud by identifying economic fraud risks during processing. However, NMFS inspects approximately one-third of the seafood consumed in the United States. FDA directs its field staff to minimize work on economic fraud issues because it considers food safety a higher priority than economic fraud. Nonetheless, FDA's health and safety actions, such as examinations of seafood imports, sometimes uncover seafood fraud incidentally. FDA's primary regulatory program for domestic seafood processors—HACCP—does not address the economic fraud risks also associated with processing. In addition, while FDA maintains a list of scientific and market names of seafood that is intended to help the seafood industry comply with FDA's regulations on product mislabeling, until January 2009, FDA had not fully updated the publicly available list it created in 1993, despite having made numerous changes since then. Finally, the guidance FDA provides to seafood processors to help ensure that their seafood products are safe does not reflect the seafood labeling requirements in the Food Allergen Labeling and Consumer Protection Act of 2004, which could incidentally help detect and prevent species substitution, since processors would need to verify the species of fish or shellfish to ensure accurate labeling.

CBP Takes a Variety of Actions to Detect Schemes to Falsify Import Information and Thereby Avoid Paying Import Duties

For imported seafood to enter U.S. commerce, the importer must file for entry with CBP and submit electronic or paper entry documents to CBP. The entry documents include basic information about the imported product, such as its type, quantity, and value. As the first step, port officials select some of these entry documents to review to determine whether to allow the imported product, including seafood, to enter U.S. commerce.[10] However, due to the large number of goods imported into the United States, port officials can only examine a fraction of incoming shipments. For example, from fiscal years 2004 to 2008, CBP officials examined between approximately 1.0 to 2.4 percent of all seafood imports. Importers must file additional documents (known as "entry summary" documents) and pay the appropriate duties, taxes, and fees on imported merchandise for consumption, including seafood products, within 10 days after CBP releases them from its custody. CBP's import specialists and other port officials select some entry summary documents to review to ensure accurate duty collection, which includes some seafood. For example, Chinese shrimp and crawfish and Vietnamese frozen fish fillets are subject to antidumping duties. On the basis of these reviews, CBP officials select importers for further review whose documentation indicates that they may be trying to avoid paying the appropriate duties.

In addition, CBP operates a national statistical sampling program, known as the Compliance Measurement Program, which randomly selects shipments of imports by commodity for review or examination to determine the degree to which they comply with customs trade laws and regulations, among other things. The Compliance Measurement Program was designed to identify trade problems or patterns of deception for specific commodities so that CBP officials can then focus their efforts against these illegal or fraudulent activities. Port officials only review a limited number of seafood entries under the program. For example, in fiscal year 2007, import specialists from all ports in the United States examined 766 seafood product entries out of 390,799 such entries and found a high compliance rate, 97.3 percent, for the applicable trade revenue laws. Although the Compliance Measurement Program could uncover seafood fraud or duty evasion, CBP officials noted that in-depth investigative work may be needed to uncover schemes to willfully defraud the U.S. government, which is beyond the scope of the examinations conducted under the program.

Another step CBP takes to detect and prevent seafood fraud is to target shipments that CBP officials suspect are part of a scheme to evade customs duties. CBP has five National Targeting and Analysis Groups (NTAG) that develop criteria to target potentially fraudulent imports. One of these NTAGs develops criteria to target potentially fraudulent shipments of seafood and reviews leads from other CBP officials and external organizations, such as trade associations, on transshipping schemes to avoid paying antidumping and countervailing duties. This NTAG researches and monitors trade trends to identify changes or patterns in trade that may signal potential fraudulent activity. For example, as part of their 2005 inquiry into an allegation of illegal transshipment of Chinese shrimp through Indonesia, the NTAG staff reviewed information on the shippers of Indonesian shrimp before and after the antidumping duty order for Chinese shrimp was put in place. They found a sharp decrease in shrimp imports from China after the antidumping duty order was issued in early 2005 and a concurrent increase in shrimp imports from Indonesia, among other countries. The NTAG staff enlisted the support of ICE to investigate Indonesian shrimp exporters who they suspected were illegally transshipping Chinese shrimp. They found that some Indonesian firms were importing Chinese shrimp and then shipping them to the United States labeled as Indonesian shrimp. CBP found that, in 2005, approximately $6 million worth of Chinese shrimp had been illegally transshipped through Indonesia to avoid antidumping duties.

While the illegal transshipment of Chinese shrimp continued through a different transshipping point, this time it also had a health- and foodsafety-related effect. In June 2007, FDA announced a countrywide import alert on five Chinese-farmed seafood products, including shrimp. This import alert required that all Chinese shrimp be detained and refused entry, unless the importer could prove the absence of unapproved drugs in the shrimp. On the basis of industry information and CBP and ICE investigations, CBP determined that Chinese shrimp was being transshipped to the United States through Malaysia. Due to this illegal transshipment, importers of Chinese shrimp were able to circumvent not only the 2005 antidumping duty but also FDA's recent import alert. In September 2007, CBP tested shipments of suspected Chinese shrimp illegally transshipped through Malaysia for the presence of unapproved drugs and found some contaminated shrimp. On the basis of CBP's information, in March 2008, FDA issued a new import alert requiring importers of shrimp from one Malaysian manufacturer to prove the absence of unapproved drugs prior to entering future shipments of shrimp into U.S. commerce.

In 2007, the NTAG that works on seafood fraud issues also helped identify another scheme importers were using in their attempt to evade antidumping duties on Chinese shrimp. Under this scheme, importers provided CBP with fraudulent information on the product type to evade antidumping duties. A precursor to breaded shrimp called "dusted shrimp" was exempted by the Department of Commerce from the antidumping duty order on imported Chinese shrimp.[11] On the basis of allegations from the U.S. shrimp industry, CBP initiated an intensive examination and sampling operation to determine whether importers were bringing in shipments of falsely declared dusted shrimp to avoid the antidumping duties on Chinese shrimp. Over the course of a 90-day period, CBP found that of the 81 alleged dusted shrimp entries examined and sampled, approximately 64 percent of the shipments did not meet the criteria to qualify as dusted shrimp. The potential loss of trade revenue from these fraudulent dusted shrimp shipments was approximately $5 million. Extrapolating back to when the antidumping duty order first became effective in 2005, CBP concluded that the importers caught importing these fraudulent dusted shrimp imported approximately $117 million worth of potentially fraudulent dusted shrimp with a possible loss of trade revenue from the uncollected antidumping duties of $132 million.

CBP's Office of Regulatory Audit provides additional support with suspected transshipping incidents to determine whether importers are participating in schemes to evade duties. CBP audits importers to ensure that they have reported and paid all trade revenue, such as antidumping duties, as required under trade laws and agreements. They perform two types of audits—a quick-response audit, which focuses on a specific issue, and a focused assessment audit, which evaluates all of a company's CBP activities and controls. The quick-response audits are focused on detecting fraudulent practices, such as unlawful transshipment of seafood to evade antidumping duties. For example, a quick-response audit concluded in 2007 found that an importer did not pay approximately $2.2 million in antidumping duties on imported Chinese shrimp that was transshipped through Indonesia. The focused assessment audits are comprehensive audits where the auditors review and test the company's management oversight processes or "internal controls" to identify areas of uncollected trade revenue, such as unpaid antidumping duties. The focused assessment audits have included seafood importers, but they have only uncovered one violation since 2005 that was related to seafood fraud.

Finally, in the event CBP identifies violations of laws, it can assess penalties against an importer. The maximum penalty amount that CBP can assess varies, depending upon whether the perpetrator's actions were fraudulent, negligent, or grossly negligent. Penalties can range from two to four times the loss of lawful duties, taxes, and fees to the U.S. government or the domestic value of the merchandise. In addition, CBP may cooperate with other agencies to pursue criminal charges against perpetrators of seafood fraud. For example, in July 2003, CBP cooperated with ICE to investigate a case where, according to ICE officials, one exporter and several importers conspired to evade antidumping duties in excess of $3 million on freshwater crawfish. According to these officials, the coconspirators were indicted for smuggling and conspiracy in November 2003, and one defendant was convicted of conspiracy in 2004.

NMFS's Voluntary, Fee-for-Service Seafood Inspection Program Helps Industry Detect and Prevent Seafood Fraud

NMFS offers two types of services to help industry address seafood fraud. NMFS provides these voluntary, fee-for-service inspection services through its Seafood Inspection Program. This program currently serves approximately 375 domestic seafood firms and 63 foreign seafood firms, and, according to senior officials in NMFS's Seafood Inspection Program, NMFS inspects approximately one-third of the seafood consumed in the United States. First, NMFS inspects shipments (known as "lots") of seafood products purchased by its clients that include retailers such as supermarkets. During lot inspections, inspectors take a random, representative sample from the seafood lot and may perform several tests on this sample, including weighing it to ensure that the actual weight matches the labeled package weight. NMFS inspectors have identified instances of seafood fraud, especially short-weighting, in seafood products processed domestically and in foreign facilities. NMFS also may be able to uncover species substitution during a lot inspection in two ways: (1) according to senior officials in NMFS's Seafood Inspection Program, NMFS inspectors are trained to visually differentiate between different types and species of fish and (2) inspectors can use the capabilities of NOAA's National Seafood Inspection Laboratory to test for and identify the species of seafood. However, NMFS does not maintain a comprehensive list of all lot inspections conducted and, thus, does not have the ability to determine the frequency with which it uncovers fraudulent seafood products.

Second, NMFS also offers a Quality Management Program that can help seafood processors decrease the potential for fraudulent seafood by applying HACCP principles to both food safety and economic fraud risks. In 2007, NMFS inspected 202 domestic seafood companies under its Quality Management Program. As part of this program, NMFS inspects seafood processors to ensure their compliance with FDA's seafood HACCP regulations that require processors to identify potential food safety hazards during processing and establish controls to mitigate them. NMFS also goes a step further and requires that seafood processors that choose to participate in the Quality Management Program (1) create a "Defect Action Plan," which identifies potential economic fraud risks during the processing of seafood, such as short-weighting, and (2) develop measures to mitigate those fraud risks. For example, the plan could require that samples of the finished seafood product be removed from the processing line every hour and the weight checked to ensure that the actual weight and the weight as stated on the package are the same. If any short-weights were identified, then all products since the last check would be weighed again, and the content amounts would be corrected. Also, the scales used in the processing phase would be checked and recalibrated, if necessary. In addition, processors can inspect seafood products upon receipt and compare the processors' species identification with the species listed on the invoices. Through these measures, processors can provide some assurance against short-weighting and species substitution.

FDA Considers Economic Issues a Low Priority Compared with Health and Safety Issues, but It Has Incidentally Identified Instances of Seafood Fraud

FDA is responsible for ensuring that foods are safe, wholesome, sanitary, and properly labeled. To that end, FDA is authorized to issue regulations to enforce the Federal Food, Drug, and Cosmetic Act. The act prohibits the misbranding or adulteration of food products, which would include seafood products that have been mislabeled, substituted, or over-glazed. FDA considers detecting violations like these a low priority and devotes minimal resources to such work, according to published program guidance and senior FDA officials. For example, FDA's program guidance to its field staff on imported and domestic seafood products states that no resources have been allocated for seafood fraud-related work, and that resource expenditures in this area should be kept to a minimum. According to FDA senior officials, the agency does not have the staff or resources to address economic fraud in addition to their food safety responsibilities, which they believe are a top priority and more important.

Nevertheless, in the course of their work examining the safety of seafood imports, FDA inspectors may uncover seafood fraud. FDA inspectors review import entry information, such as the type of seafood being imported and the importer's history of violations, if any; physically examine the imported goods; and collect samples for laboratory analysis to identify those that are potentially in violation of U.S. food regulations and laws. FDA may refuse to allow an imported seafood product to enter U.S. commerce if it appears to be adulterated or misbranded or for other violations of regulations and laws that

FDA administers. However, from fiscal years 2003 to 2008, only 1 percent of the refusals of imported seafood products were related to seafood fraud.

FDA also maintains an "import alert" list to detain entries of imported foods that appear to have significant recurring violations. FDA currently has three import alerts related to seafood fraud. One import alert lists 10 foreign firms that were found to have declared an assortment of seafood products under the name of a fictitious, incorrect, or substituted species. The other two import alerts provide information to FDA's field staff about the potential mislabeling of two specific types of fish—catfish and red snapper. The import alerts inform field office staff of the species of fish that can legally be labeled as catfish or red snapper and guidance on how to handle incoming shipments that may not be accurately labeled. However, FDA officials only physically examine a small percentage of imported seafood—about 2 percent of all seafood entries from fiscal years 2003 to 2008. Of the 2 percent of imported seafood examined by FDA officials, approximately 0.05 percent of these examinations were related to seafood fraud.

FDA's primary oversight program for domestic seafood firms is the HACCP program, but the focus of this program is health and safety not economic fraud. FDA's seafood HACCP regulations require seafood processors to identify and establish controls to mitigate potential food safety hazards. FDA inspects domestic firms involved in the production, storage, and distribution of fish and fishery products to ensure that their HACCP plans are properly designed and implemented. However, FDA inspectors spend very little time looking for seafood fraud. For example, the percentage of domestic seafood firm inspections where investigators conducted seafood fraud work was approximately 0.5 percent from fiscal years 2003 to 2008. FDA also inspects some importers and foreign processors for HACCP compliance because, as the agency noted in publishing its final HACCP regulations, the importer and foreign processor share responsibility in complying with importing regulations. Importers are required either to (1) obtain seafood products from a country that has an active memorandum of understanding or similar agreement with FDA that documents the equivalency or compliance of the foreign inspection system with the U.S. system or (2) implement written verification procedures to show that the foreign processor has complied with the HACCP regulations. Furthermore, FDA inspects foreign fish and fishery processors that export seafood to the United States, but this number of inspections is very limited. For example, in fiscal year 2007, FDA inspected only 61 of 14,569 registered foreign seafood firms.

In its 1994 seafood HACCP proposed rule, FDA recommended that seafood firms use HACCP-like measures to control for economic fraud because, as FDA stated, seafood fraud could also impact food safety. In the proposed rule, FDA stated that the " . . . misidentification of species may also have adverse public health consequences."[12] The connection between economic fraud and food safety can be seen in an incident in 2007 where two individuals became ill after eating puffer fish. According to FDA, the puffer fish was imported into the country mislabeled as monkfish. Unlike monkfish, puffer fish contains a potentially deadly toxin called *tetrodotoxin*. CBP and ICE's shrimp transshipping investigation also highlights the connection between economic fraud and food safety. CBP and ICE's investigation found that foreign manufacturers and importers were not only attempting to circumvent antidumping duties by sending Chinese shrimp to the United States through Malaysia, but these companies were also evading an FDA import alert aimed at stopping adulterated Chinese shrimp from entering the United States. Incidents like these and others have led FDA officials, including a senior official in FDA's Office of Food Safety, to reconsider the agency's limited level of effort regarding economic fraud because they now believe that, if left unregulated, over time, seafood fraud may create food safety problems. Specifically, that same official said that it may be time for FDA to reconsider incorporating nonfood safety hazards, such as economic fraud, in the seafood HACCP regulations. According to senior FDA officials, the agency can still take actions, such as reviewing labels and issuing import alerts to prevent seafood fraud. However, these officials acknowledged that requiring measures in HACCP to address seafood fraud would build prevention into the processing of seafood. In addition, NMFS officials that administer the agency's Quality Management Program believe that amending FDA's HACCP regulations to include measures to address economic fraud risks would require minimal resources for the seafood industry and FDA to implement. Moreover, according to officials in the Canadian Food Inspection Agency, Canada has a program similar to NMFS's program that requires seafood processors to identify and mitigate both food safety hazards and economic fraud risks. However, because FDA inspectors primarily inspect for compliance with HACCP regulations, until the agency amends the HACCP regulations to include measures to identify and mitigate economic fraud risks, its inspectors will continue to spend limited time ensuring against mislabeled products, and seafood processors may be less attentive to protecting against fraud.

Another opportunity for FDA officials to detect seafood fraud is during their review of food labels. According to officials in FDA's Division of Import

Operations and Policy, food label reviews sometimes occur during HACCP inspections of domestic seafood processors and examinations of imported seafood products. According to FDA guidance, investigators are to review labels on at least three food products when conducting a domestic or foreign firm inspection. The review entails ensuring that the label complies with all relevant food label laws and FDA regulations. For example, the Federal Food, Drug, and Cosmetic Act, as amended, prohibits the "misbranding" of food, which includes, among other things, labeling that is false or misleading. In addition, food labels must generally include an ingredients list that identifies the product's ingredients by their common or usual names in order of predominance by weight. Food labels must also accurately state the contents in terms of weight, measure, or numerical count of the product. However, our September 2008 report found that FDA does not have reliable data on the number of labels reviewed, nor does it track the complete and timely correction of the violations it identifies during the reviews.[13] As a result, FDA cannot provide reliable information on the number of label reviews that identified seafood fraud, nor on any corrections of the seafood fraud violations it identified.

FDA may also detect seafood fraud through complaints from industry associations. For example, on several occasions during 2008, NFI provided FDA with copies of public, written solicitations from foreign and domestic companies offering to sell packages of seafood packed to 80 or 90 percent of the labeled package weight for a discount. These short-weighted products could then potentially be fraudulently sold in U.S. commerce for the full price on the basis of the labeled package weight. According to an FDA Consumer Safety Officer, FDA wanted NFI to provide additional information on these companies, such as their locations, because some of these companies were not in FDA's inventory of seafood processors or importers, and FDA could not identify them with certainty. However, we found that some of the companies were listed in FDA's inventory because FDA had previously inspected them. According to a senior FDA official, FDA is still working with NFI on this matter; however, as of December 2008, the agency had taken no action against any of the companies that sent solicitations.

In addition, FDA may detect seafood fraud from consumers and others who contact FDA field staff, by telephone, to complain about issues such as suspected short-weighted seafood products or species substitution. For example, in one case, a consumer complained to FDA about frozen shrimp mislabeled as a product of Mexico when a second label underneath the first indicated that it was a product of Thailand. Typically, the FDA District

Consumer Complaint Coordinator would field the telephone call, collect information from the consumer, such as the description of the product and the problem with the product, and would attempt to determine the responsible seafood firm. The Complaint Coordinator or other FDA field staff would evaluate the complaint and determine whether immediate action is required or whether the information would be used for surveillance in the next inspection of the seafood firm. However, our 2008 food labeling report found that FDA program managers could not use FDA's data system that captures consumer complaints to track their timely and appropriate resolution because these data were not entered into the system in a manner that would facilitate analysis. Specifically, standard terminology was not used and information on complaint resolutions was captured in different data fields.[14] We found similar issues regarding the follow-up of consumers' complaints of seafood fraud. We reviewed FDA summary data for 105 consumer complaints related to such things as misbranded, mislabeled, and short-weighted seafood products from fiscal years 2004 through 2007. We judgmentally selected and reviewed FDA's internal documents for 5 of 105 consumer complaints and found that for all 5 of the complaints, FDA indicated that it would follow up on the complaints during the next inspection of the firm that had produced the seafood product in question. However, we found no information in FDA's inspection reports to indicate that FDA had followed up on three complaints. There was no inspection report available for the fourth complaint, and for the fifth complaint, there was a report for an inspection that was conducted a few days prior to the filing of the complaint.

There are several actions that FDA can take if seafood fraud is discovered during an inspection of a seafood firm. FDA may issue a warning letter—which is a notice that enforcement actions may be forthcoming if corrections are not made—to firms for serious violations of regulatory significance. For less serious violations, FDA may send an untitled letter, which is an informal communication that corrective actions are needed. However, FDA issued no warning letters or untitled letters regarding seafood fraud from fiscal years 2005 through 2008. Additionally, FDA may initiate enforcement actions against seafood firms, such as seizing and removing seafood from the marketplace; obtaining an injunction to stop a company from engaging in behavior that violates certain prohibitions of the Federal Food, Drug, and Cosmetic Act and regulations; or barring seafood firms from importing goods into the United States. FDA, however, has not taken any of these actions for seafood fraud violations since 2000, according to a senior FDA official. However, FDA's Office of Criminal Investigations has investigated allegations of seafood fraud and provided information to the Department of Justice for legal action against the perpetrators.

FDA helps the seafood industry avoid species substitution by maintaining on its agency Web site a publicly available "Seafood List," which is a compilation that includes the scientific and market names for imported and domestic seafood. The Seafood List is intended to promote uniformity in the use of FDA-acceptable market names by the seafood industry and to provide consistent advice on these names. For example, "catfish" can only be used as a market name for fish that belong to the family *Ictaluridae*, even though there are other species of fish whose vernacular name may include the word catfish. However, before January 2009, the public version of the Seafood List had not been kept up to date. According to an FDA official responsible for the Seafood List, even though over 400 changes have been made to FDA's internal version of the list since 1993, only about 22 changes had been made to the public version of the list as of the end of 2008. However, in January 2009, FDA made the revised and fully updated list available to the public on its Web site.

Other countries, such as Canada and Australia, maintain a similar type of Seafood List. However, their policies and procedures for maintaining these lists differ from those of FDA. For example, Canada publishes on the Internet guidelines and criteria that it uses to assign a new name for a species of seafood. In addition, the Canadian government will not allow a new name for a fish if it is similar or resembles the name of a fish with a higher market value. Also, changes made to Canada's seafood list are disseminated to an e-mail distribution list that interested parties can join. In contrast, until January 2009, FDA did not provide information to the public on the guidelines or criteria that it used to determine acceptable market names for seafood. In addition, stakeholders do not have the opportunity to comment on proposed changes to the Seafood List, unless the change was required by law or regulation. In contrast, Australia allows for a 3-month public consultation period prior to finalizing the change. A senior FDA official told us that stakeholders can write to FDA at any time with comments about the Seafood List, but that there is no formal comment period before changes are made to the list because it is considered guidance, not an agency regulation.

A recent example of a change to the Seafood List illustrates some of these issues. In April 2005, FDA allowed a restaurant to advertise langostino—with the common names "squat lobster" and "Colorado langostino"—as "langostino lobster." However, as of December 2008, the publicly available Seafood List had not been updated to reflect this change. According to the list on December 10, 2008, the market name for langostino is "langostino," and the common names are "squat lobster" or "Colorado langostino." Furthermore, according to officials from the Maine Lobster Promotion Council, the council wrote to FDA to protest this change to the naming of langostinos because it allows a lower-market-value seafood, langostinos, to take advantage of the higher market value and reputation of the American lobster name.[15] According to officials from the Maine Lobster Promotion Council, FDA did not respond to its inquiries regarding the reasons for the name change for langostinos. An FDA official responsible for the Seafood List told us that FDA responded to the council by e-mail in August 2008 stating that, "we have not objected to use of the term langostino lobster for various species of squat lobsters."

According to a senior FDA official, seafood labeling requirements from the Food Allergen Labeling and Consumer Protection Act of 2004 may help prevent species substitution, but FDA has not updated its guidance to the seafood industry to fully reflect the act's labeling requirements. The act requires the product label to contain the name of a food source from which a major food allergen is derived, and, when the major food allergen is fish or Crustacean shellfish, the specific species of fish or shellfish must be listed to notify consumers with food allergies of a particular type of fish species. According to the same FDA official, the act's labeling requirement may help detect and prevent species substitution, since processors would need to verify the type of fish or shellfish they are processing to ensure accurate labeling. However, these seafood labeling requirements are not reflected in FDA's guidance to the seafood industry on the development of their HACCP plans to prevent and control the health and food safety hazards associated with seafood—the *Fish and Fisheries Products Hazards and Controls Guidance*.[16] According to a senior FDA official, seafood processors should be aware of the act's requirements, and FDA can still enforce those requirements, even if they are not reflected in the guidance. Even so, senior FDA officials told us that they plan to update the guidance to reflect the Food Allergen Labeling and Consumer Protection Act of 2004 labeling requirements, but they could not provide us with a publication time frame.

LIMITED COLLABORATION EXISTS AMONG THE KEY AGENCIES RESPONSIBLE FOR DETECTING AND PREVENTING SEAFOOD FRAUD

Limited collaboration among CBP, NMFS, and FDA to detect and prevent seafood fraud can diminish their efficiency and effectiveness in dealing with this issue. CBP, ICE, FDA and its Office of Criminal Investigations, and NOAA's Office for Law Enforcement have worked together on criminal cases against individuals and companies suspected of committing seafood fraud. CBP, ICE, and FDA are working together to target commodities that pose health and safety risks, and NMFS and FDA have worked on joint laboratory efforts to share standards for seafood species identification. However, CBP, NMFS, and FDA have not established common goals related directly to detecting and preventing seafood fraud or joint strategies to help achieve the goals. In addition, these key agencies have not established policies and procedures to promote effective collaboration and better leverage resources to achieve their common goal. As a result, the agencies have not taken advantage of opportunities to share information that could benefit each agency's efforts to detect and prevent seafood fraud, nor have they identified similar and sometimes overlapping activities that could be better coordinated to use limited resources more efficiently and effectively.

CBP, NMFS, and FDA Have Not Taken Any of the Key Steps to Develop an Enhanced Collaborative Working Relationship Related to Detecting and Preventing Seafood Fraud

CBP, ICE, FDA and its Office of Criminal Investigations, and NOAA's Office for Law Enforcement have worked together on developing and investigating criminal cases against individuals and companies suspected of committing seafood fraud. For example, in one situation, ICE and the Office for Law Enforcement developed a case against an individual and his companies in Florida, who, aided by exporters in Vietnam, imported thousands of pounds of catfish into the United States labeled as grouper and other fish to avoid paying antidumping duties. The case resulted in an indictment and a guilty plea by the defendant. In addition, CBP, ICE, and FDA are working together to target commodities that pose health and safety risks through Operation Guardian, which is an enforcement initiative to deal with imported substandard, tainted, and counterfeit products. Operation Guardian's efforts have led to seizures of such commodities as pharmaceuticals, steel components, honey, shrimp, and toys.

However, despite CBP, NMFS, and FDA having responsibilities related to seafood fraud detection and prevention, these agencies have not implemented key practices our previous work has identified that can enhance and sustain collaboration among federal agencies.[17] Specifically, CBP, NMFS, and FDA have not worked with each other to (1) identify their shared individual goals related to seafood fraud detection and prevention and (2) develop a common overarching goal or goals. Because these agencies have not identified a common goal, they also have not implemented the next practices we identified for effective collaboration and established joint strategies to achieve that goal or agreed on their respective roles and responsibilities in achieving the shared goal. The agencies also have not identified ways to maximize and leverage their resources by agreeing on the resources and activities each agency can commit to accomplishing the common goal, nor have they established procedures and policies for working together systematically across agency lines. Because the agencies have not implemented any of the key practices that enhance collaboration, they have missed opportunities to share data among themselves that could enhance seafood fraud detection and prevention and leverage resources. They also have not identified similar or overlapping activities that could be better coordinated or consolidated so that limited funds would be used more efficiently and effectively.

CBP, NMFS, and FDA Are Missing Opportunities to Share Information to Better Detect and Prevent Seafood Fraud

CBP, NMFS, and FDA each collect information on seafood products to meet their respective responsibilities, but they do not always share information that could be used to detect and prevent seafood fraud. CBP collects information on seafood imports, such as product type, product quantity, and country of origin, through the review and examination of imported goods and import entry documents, audits, and laboratory analysis. NMFS collects information in lot inspection reports that identify short-weighted domestic and imported products. FDA collects information on imported seafood products, such as the accuracy of product labeling, through entry document reviews, food label reviews, product examinations, inspections, and laboratory analysis. Some of the information these agencies collect could be used to identify seafood fraud, such as the names of the importers and the seafood products that were illegally transshipped to avoid customs duties or were mislabeled. However, these agencies have not developed procedures to identify or share useful information. For example, CBP and FDA both find seafood imports with inaccurate product type, weight, and country-of-origin labels. By sharing this information, CBP and FDA would have more data about violative imported products and could target those products with the most violations or greater likelihood to be violative for more frequent inspections. Additionally, CBP and FDA could compare information each agency receives from importers on the product type and country of origin to determine if this information is consistent, which would help ensure that importers were not attempting to circumvent an antidumping duty or an FDA import alert. An official in FDA's Division of Import Operations and Policy told us that while there are no formal mechanisms for sharing this information between FDA and CBP, FDA makes referrals regarding violative imported products to CBP on an ad hoc basis.

In addition to information on seafood products, CBP, NMFS, and FDA also have information on seafood importers that each collects through audits and inspections, respectively, but do not share this information with each other. During an importer audit, CBP ensures that importers have reported and paid all trade duties as required under trade laws and agreements on the basis of the product documentation and labeling. During an inspection of a seafood importer, FDA may review importer documentation and collect samples to identify products not in compliance with the Federal Food, Drug, and Cosmetic Act's labeling requirements. FDA and CBP may both identify discrepancies in product information supplied by seafood importers, such as declaring incorrect information on import documents or having inadequate proof of documentation, when checking for compliance with labeling requirements. Because CBP and FDA do not share the results of their respective audits and inspections, they do not have the most comprehensive information on noncompliant seafood importers that could help identify those with a greater likelihood of problems in the future. In addition, NMFS does not share information with FDA from its lot inspections about importers and processors that produced short-weighted seafood products on a regular basis, so FDA could further investigate this concern during future examinations or inspections.

In investigating illegal transshipment or other schemes to avoid duties, CBP and ICE may require information on foreign seafood producers critical to their investigations. However, CBP and ICE may be denied access to the countries where the problematic seafood producers are located. For example, in 2007, CBP and ICE were denied entry into a country to investigate an alleged transshipping scheme to avoid paying antidumping duties and, thus, could not collect crucial information about the foreign seafood producers in question. CBP could potentially benefit from information obtained from FDA and NMFS, both of which conduct inspections of foreign seafood producers.[18] According to senior CBP and ICE officials, they would find information from FDA's and NMFS's foreign inspections beneficial, depending on the facts of particular enforcement activities they are pursuing.

Lack of Collaboration by CBP, NMFS, and FDA to Detect and Prevent Seafood Fraud Has Resulted in Overlapping Actions and Inefficient Use of Resources

Not only does the lack of collaboration create inefficient information sharing between the key federal agencies, it also creates overlapping agency efforts and inefficient use of government resources. NMFS and FDA have similar inspection programs—NMFS inspects facilities, on request, for health, safety, and economic integrity issues, while FDA focuses its inspections on health and safety concerns. However, an FDA official said that the agency is not sure whether it can rely on NMFS inspections, in part due to concerns about potential conflicts of interest because NMFS is paid by industry to conduct its inspections. FDA has identified these potential conflicts as an impediment to fully using NMFS inspection efforts in the past. In our 2004 report on FDA's imported seafood safety program, we stated that an official raised concerns about potential conflicts of interest with NMFS inspections, but that other officials thought that these concerns could be addressed in an agreement between the two agencies.[19] We recommended that FDA and NMFS develop a memorandum of understanding (MOU) that, in part, would use and leverage NMFS inspection services to more efficiently and effectively monitor the safety of imported seafood. In response, FDA stated that there were already three MOUs between FDA and NMFS that dealt with seafood safety and inspection operations, but that it would explore additional opportunities to better leverage NMFS inspection resources and more efficiently and effectively protect the public health. Among the three MOUs, the 1974 MOU between FDA and NMFS stated, in part, that NMFS would provide FDA with information on establishments under contract with it, and that such inspections and consultations with FDA should diminish the need for

FDA inspections. Despite FDA's statements and the provisions in its 1974 MOU, FDA still does not take into account whether NMFS has already inspected a facility when FDA determines which facilities it will inspect. For example, from 2005 through 2008, NMFS inspected one facility we visited at least four times a year, yet FDA also inspected it in 2005, 2006, and 2008. Furthermore, neither agency found any significant issues during their inspections of this facility. Overall, in fiscal year 2007, FDA inspected 120 facilities that were also inspected by NMFS, while FDA had not inspected 1,464 other facilities since before fiscal year 2003. Also during fiscal year 2007, NMFS inspected 88 facilities that FDA either had not inspected within the same fiscal year or had not inspected at all. In its technical comments to our draft report, FDA stated that it is currently negotiating an MOU with NMFS that is intended to address its concerns about potential conflicts of interest.

Furthermore, CBP, NMFS, and FDA each have their own laboratory capabilities for, among other things, determining the species of seafood samples they receive. Moreover, because these agencies use different testing methodologies and standards for species identification, they do not acknowledge each others' laboratory results, nor do they share the species standards they have developed. CBP uses DNA sequencing to identify a seafood species, and FDA and NMFS use the isoelectrophoresis method.[20] While CBP has developed some authenticated DNA samples of fish species, it uses GenBank DNA sequences as a guide to conduct most of its laboratory testing for species identification.[21] FDA and NMFS do not believe that the GenBank data are sufficiently accurate to use as the basis for a regulatory action, such as providing validated evidentiary support to prove species substitution during a criminal prosecution. As a result, FDA is in the process of developing its own secure database of DNA sequences for seafood species identification. FDA laboratory officials told us the agency has developed DNA sequences for 72 seafood species and is in the process of adding about 100 more species. NMFS has its own library of standards for seafood species identification but has had meetings with FDA regarding sharing species standards and getting DNA analysis approved by an international accreditation agency as an official method for species identification. According to FDA laboratory officials, the agency would be willing to share its species standards with CBP as well, but neither agency has had any discussions on this matter. According to a senior CBP laboratory official, CBP has no need to access any standards other than those contained in GenBank. Both CBP and FDA could increase the number of DNA sequences they have available to identify species substitution by combining and sharing their standards. However, CBP, NMFS, and FDA have not collaborated with each other to develop a comprehensive DNA database.

CONCLUSIONS

Although FDA has tools that it could use to help detect and prevent seafood fraud, the agency is not using them as efficiently and effectively as it could. For example, the agency's primary regulatory program for seafood firms—HACCP—does not require firms to identify and establish controls to mitigate economic fraud risks, only potential food safety hazards. Also, the agency does not provide the opportunity for stakeholder comments prior to making changes to the Seafood List, and it has not routinely updated the public version of the list whenever FDA makes changes to it. This has limited the effectiveness of the list in promoting uniformity in seafood species names and in preventing species substitution. Likewise, not updating the *Fish and Fisheries Products Hazards and Controls Guidance* to reflect the labeling requirements of the Food Allergen Labeling and Consumer Protection Act of 2004 has limited the usefulness of the guidance in preventing species substitution.

Because responsibilities for detecting and preventing seafood fraud are shared among three key federal agencies—CBP, NMFS, and FDA— effective collaboration is important to operating efficiently and effectively and to producing a greater public benefit than if the agencies acted alone. Currently, however, the three agencies are mostly acting alone and have not implemented key practices to begin collaborating more efficiently and effectively, such as identifying a common goal and establishing joint strategies to achieve the goal. As a result, the three agencies are not sharing information that could be used to detect or prevent seafood fraud and are working on creating individual databases of seafood species standards and using different methods to do so. Until these agencies begin collaborating more efficiently and effectively to detect and prevent seafood fraud, duplication of efforts may continue to occur, and the agencies may continue to miss opportunities to use limited federal funds more efficiently and effectively.

RECOMMENDATIONS FOR EXECUTIVE ACTION

To help reduce the prevalence of seafood fraud and improve FDA's actions to detect and prevent seafood fraud, we are recommending that the Commissioner of the Food and Drug Administration take the following three actions:

- propose amendments to FDA's seafood HACCP regulations to include requirements that covered facilities include control points that can be used to identify and mitigate economic fraud risks;
- provide the opportunity for stakeholder comments prior to formalizing any changes to the Seafood List not required by law or regulation and routinely update the public version of the list whenever FDA makes any changes; and
- update the *Fish and Fisheries Products Hazards and Controls Guidance* to reflect the seafood labeling requirements of the Food Allergen Labeling and Consumer Protection Act of 2004.

  To maximize the efficiency and effectiveness of each agency's efforts to detect and prevent seafood fraud and to increase interagency collaboration, improve information sharing, and reduce overlaps, we recommend that the Commissioner of Customs and Border Protection, the Under Secretary of Commerce for Oceans and Atmosphere, and the Commissioner of the Food and Drug Administration take the following two actions:
- develop goals, strategies, and mechanisms to share information and resources related to seafood fraud detection and prevention across agency boundaries and
- create a federal agencywide library of seafood species standards.

APPENDIX I. SCOPE AND METHODOLOGY

The three federal agencies that play key roles in detecting and preventing seafood fraud are (1) the Department of Homeland Security's Customs and Border Protection (CBP), (2) the Department of Commerce's National Marine Fisheries Service (NMFS), and (3) the Department of Health and Human Services' Food and Drug Administration (FDA). We reviewed data and documents from each agency on the amount and nature of seafood fraud that they have identified and any corrective actions they have taken against fraud perpetrators. We also reviewed program guidance, inspection operation manuals, and other documentation and interviewed knowledgeable officials to determine each agency's overall approach and specific actions to detect and prevent seafood fraud—including the priority given to seafood fraud detection and prevention among their other responsibilities. We also observed FDA's and NMFS's inspections of seafood processing facilities. In addition, we obtained examples of criminal investigations of seafood fraud to understand the nature and impact of seafood fraud.

For this chapter, we relied on the findings of our September 2008 report to assess the reliability of FDA's data.[22] The data we used came from the same data sources used in our 2008 report, in which we found the FDA data to be sufficiently reliable for the purposes of this chapter. To assess the reliability of CBP and NMFS data, we performed testing for obvious errors in accuracy and completeness and reviewed relevant documentation. We also worked closely with agency officials to identify and resolve any data problems. We determined that the data from these three agencies were sufficiently reliable for the purposes of this chapter.

Finally, we also met with representatives from six major seafood industry associations to determine their views on the nature of the seafood fraud problem, the actions they have taken to help prevent seafood fraud, and how they interact with the key federal agencies. We interviewed officials from the Catfish Farmers of America, Maine Lobster Promotion Council, National Fisheries Institute, National Restaurant Association, Southeastern Fisheries Association, and Southern Shrimp Alliance.

We visited the Boston, New York/Newark, and Los Angeles/Long Beach ports of entry. During these visits, we observed CBP's process for reviewing import entry documents, discussed local targeting efforts and whether any collaborative efforts related to seafood fraud existed between CBP and FDA, and toured the port facilities. At the Los Angeles/Long Beach ports, we also met with FDA officials to discuss and observe their process for inspecting imported seafood.

We also interviewed officials from four states—California, Florida, New York, and Texas—to determine their roles, responsibilities, and interactions with the federal agencies in detecting and preventing seafood fraud. We selected these states on the basis of their population size and, because of their coastal location, their having a potential port of entry for imported seafood. Additionally, we interviewed officials at Alabama's Food and Drug

Laboratory because of its active efforts in developing laboratory methods for testing seafood products.

To determine the extent that the key federal agencies collaborate with each other to help detect and prevent seafood fraud, we first identified practices that our previous work indicated can help enhance and sustain collaboration among federal agencies.[23] For the purposes of this chapter, we focused on the first five of the eight practices we previously identified for enhancing and maintaining effective collaboration among federal agencies: (1) define and articulate a common goal; (2) establish mutually reinforcing or joint strategies to achieve that goal; (3) identify and address needs by leveraging resources; (4) agree on roles and responsibilities; and (5) establish compatible policies, procedures, and other means to operate across agency boundaries. We did not address the following practices: (1) develop mechanisms to monitor, evaluate, and report on results; (2) reinforce agency accountability for collaborative efforts through agency plans and reports; and (3) reinforce individual accountability for collaborative efforts through performance management systems. We did not address these practices because we found that CBP, NMFS, and FDA had not implemented the first five practices. As a result, because limited collaborative activities were under way, we did not expect the agencies to have developed mechanisms to monitor and report on the results of their collaboration, reinforce accountability by preparing reports, or establish performance management systems. We also analyzed agency documents and interviewed officials from CBP, NMFS, and FDA to determine the extent to which they had (1) implemented the first five previously mentioned collaboration practices, (2) identified and shared information that could be beneficial in addressing seafood fraud, and (3) engaged in overlapping seafood fraud-related activities.

We conducted this performance audit from January 2008 to February 2009, in accordance with generally accepted government auditing standards. Those standards require that we plan and perform the audit to obtain sufficient, appropriate evidence to provide a reasonable basis for our findings and conclusions based on our audit objectives. We believe that the evidence obtained provides a reasonable basis for our findings and conclusions based on our audit objectives.

APPENDIX II. COMMENTS FROM THE DEPARTMENT OF COMMERCE

UNITED STATES DEPARTMENT OF COMMERCE
The Deputy Under Secretary for
Oceans and Atmosphere
Washington, D.C. 20230

JAN 2 7 2009

Ms. Lisa Shames
Director, Natural Resources
and Environment
U.S. Government Accountability Office
441 G Street, NW
Washington, D.C. 20548

Dear Ms. Shames:

Thank you for the opportunity to review and comment on the Government Accountability Office's draft report entitled, *Seafood Fraud: FDA Program Changes and Better Collaboration among Key Federal Agencies Could Improve Detection and Prevention* (GAO-09-78). On behalf of the Department of Commerce, I enclose the National Oceanic and Atmospheric Administration's programmatic comments to the draft report.

Sincerely,

Mary M. Glackin
Deputy Under Secretary
for Oceans and Atmosphere

Enclosure

Department of Commerce
National Oceanic and Atmospheric Administration
Comments on the Draft GAO Report Entitled
"FDA Program Changes and Better Collaboration among Key Federal Agencies Could Improve Detection and Prevention"
(GAO-09-258/January 2009)

General Comments

In general, the report on seafood fraud detection does a fair and thorough job in assessing and isolating the issues and concerns shared by the named agencies, including the National Oceanic and Atmospheric Administration (NOAA). The report will be used to assist NOAA to determine more effective methods in seafood fraud detection and assist in prioritizing the concerns for correction and implementation. The report and the recommended NOAA solutions will be shared with the proper industry groups to facilitate changes in seafood fraud and its detection.

It should be noted that, while throughout the draft report much emphasis was placed on the NOAA activities relative to the Seafood Inspection Program, NOAA's Office of Law Enforcement also plays a pivotal role in seafood fraud activities when detected.

NOAA Response to GAO Recommendations

"To maximize the efficiency and effectiveness of each agency's efforts to detect and prevent seafood fraud and to increase interagency collaboration, improve information sharing, and reduce overlaps, we recommend that the Commissioner of Customs and Border Protection, the Under Secretary of Commerce for Oceans and Atmosphere, and Commissioner of the Food and Drug Administration take the following two actions:"

Recommendation 1: "develop goals, strategies, and mechanisms to share information and resources related to seafood fraud detection and prevention across agency boundaries"

NOAA Response: NOAA agrees with this recommendation. Such strategies and mechanisms have already been initiated in relation to other areas of seafood inspection and production. The NOAA Seafood Inspection Program, aligned with other NOAA agencies has recently adopted the strategy to research and implement proactive testing and evaluation methods to detect and isolate economic fraud practices to protect the consumer and assist industry in enhanced processing methods. NOAA will certainly work with the Department of Homeland Security's Customs and Border Protection and the Department of Health and Human Services' Food and Drug Administration to further this goal and to reduce seafood fraud in the United States.

Recommendation 2: "create a federal agency-wide library of seafood species standards"

NOAA Response: NOAA agrees with this recommendation. Such a library of documented seafood species standards shared among federal agencies would lead to

NOAA Response: NOAA agrees with this recommendation. Such a library of documented seafood species standards shared among federal agencies would lead to increased program efficiencies and effectiveness by reducing duplication and creating a more comprehensive collection of species standards. The National Seafood Inspection Laboratory is currently holding discussions with the other agencies on this subject and NOAA's National Marine Fisheries Service will continue to implement this recommendation using existing resources.

Appendix III. Comments from the Department of Health and Human Services

DEPARTMENT OF HEALTH & HUMAN SERVICES   OFFICE OF THE SECRETARY

Assistant Secretary for Legislation
Washington, DC 20201

JAN 3 0 2009

Lisa Shames, Director
Natural Resources and Environment
U.S. Government Accountability Office
441 G Street N.W.
Washington, DC 20548

Dear Ms. Shames:

Enclosed are comments on the U.S. Government Accountability Office's (GAO) report entitled: "Seafood Fraud: FDA Program Changes and Better Collaboration Among Key Federal Agencies Could Improve Detection and Prevention (GAO-09-258).

The Department appreciates the opportunity to review this report before its publication.

Sincerely,

Barbara Pisaro Clark

Barbara Pisaro Clark
Acting Assistant Secretary for Legislation

Attachment

FDA's General Comments to the U.S. Government Accountability Office's (GAO) Draft Report Entitled, *Seafood Fraud – FDA Program Changes and Better Collaboration Among Key Federal Agencies Could Improve Detection and Prevention (GAO-09-258)*

The Food and Drug Administration (FDA) appreciates the opportunity to review and comment on the Government Accountability Office's (GAO) draft report. GAO has raised some important issues regarding FDA's seafood program. FDA strives continually to advance its public health mission, and this includes efforts to improve the safety, sanitation, suitability, and proper labeling of seafood.

FDA's Specific Comments on GAO Recommendations

GAO Recommendation 1:

To help reduce the prevalence of seafood fraud and improve FDA's actions to detect and prevent seafood fraud, we are recommending that the Commissioner of the Food and Drug Administration propose amendments to FDA's seafood HACCP regulations to include requirements that covered facilities include control points that can be used to identify and mitigate economic fraud risks.

FDA Response:

FDA agrees with GAO that it is appropriate at this time to reassess whether to recommend that processors include nonsafety hazards. When FDA proposed the Seafood HACCP Regulation, 21 CFR 122, in 1994, FDA included a recommendation, not a requirement, that HACCP plans include controls for hazards unrelated to safety, such as economic fraud. Approximately 75 comments addressed this provision, the vast majority of which urged that it be eliminated from the regulations. Those that argued for removal stated that 1) HACCP for safety purposes would be a big challenge for both the industry and regulators, and that inclusion of nonsafety hazards might be overwhelming; 2) nonsafety hazards are covered adequately by existing FDA regulations and by industry quality control programs; 3) inclusion of nonsafety hazards deviates from internationally recognized HACCP principles; and, 4) inclusion of nonsafety hazards would dilute and jeopardize a desirable industry focus on safety. Those that argued for inclusion of economic adulteration stated that 1) the same conditions of processing that affect the occurrence of safety hazards affect the occurrence of nonsafety hazards, making the two control systems compatible; 2) an improvement in consumer confidence in seafood cannot be achieved without improvements relative to economic deception; 3) the seafood industry considers economic fraud to be the most significant hazard affecting the marketing of its products; 4) species substitution can be safety related; 5) HACCP controls would likely enhance compliance with existing nonsafety standards; and, 6) inclusion of controls for economic fraud would not significantly increase the costs to industry.

After review of the comments, FDA concluded that the HACCP system would have to mature and FDA would need to learn more before it could determine whether a mandatory HACCP program should include nonsafety matters. Additionally, the agency noted that the statutory provisions that form the basis for the Seafood HACCP regulations are safety provisions.

FDA's General Comments to the U.S. Government Accountability Office's (GAO) Draft Report Entitled, *Seafood Fraud – FDA Program Changes and Better Collaboration Among Key Federal Agencies Could Improve Detection and Prevention (GAO-09-258)*

FDA's application of HACCP is intended for the effective enforcement of sections 402(a)(1) and (a)(4) of the Federal Food, Drug, and Cosmetic Act, which apply to products that contain substances that may render the product injurious to health and to processing conditions that are insanitary and that could render a product injurious to health. Thus, the only issue for FDA to consider in the preparation of the final regulations was whether to retain the recommendation, not requirement, to include nonsafety hazards in processors' HACCP programs. FDA was persuaded by the point raised by some comments that advisory provisions of regulations are often confused with or misapplied as requirements. For these reasons, the recommendation to include economic fraud in HACCP plans was not included in the final regulation.

After more than a decade of implementation, FDA recognizes the Seafood HACCP program as a mature program, and agrees with GAO that this is an appropriate time to reassess whether to recommend that processors include nonsafety hazards, such as economic fraud, in their HACCP plans.

GAO Recommendation 2:

To help reduce the prevalence of seafood fraud and improve FDA's actions to detect and prevent seafood fraud, we are recommending that the Commissioner of the Food and Drug Administration publicize the criteria FDA uses to revise the Seafood List, provide the opportunity for stakeholder comments prior to formalizing any changes to the list not required by law or regulation, and routinely update the public version of the list whenever FDA makes any changes.

FDA response:

FDA agrees with GAO that it should reassess the mechanism to seek stakeholder comments on potential changes in the list of acceptable market names and inform stakeholders of the changes.

The last hard copy version of *The Seafood List* was published in 1993. Since then FDA has made the decision to produce the document only in an electronic version, available on the Agency website. The first and most recent electronic-only version of the document was published in 2002. Between the 1993 and 2002 version, there were market name changes associated with 22 species of fish, all mandated by law or regulation (e.g., catfish and brown king crab). Changes mandated by regulation undergo a public notice and comment period. Since 2002, FDA has accepted new market names for an additional 23 species of fish, none of them mandated by law or regulation.

FDA has recently completed work on the next version of the document and expects that it will be posted on the Agency's website very shortly. This version will include the 23 most recent market name changes (e.g., langostino lobster). It will also include approximately 400 changes in scientific names and additional species names in order to stay current with taxonomic convention. These latter changes do not affect industry's understanding of suitable market names. In addition, the guidance will include a description of the factors that FDA uses in determining acceptable market names.

FDA's General Comments to the U.S. Government Accountability Office's (GAO) Draft Report Entitled, *Seafood Fraud – FDA Program Changes and Better Collaboration Among Key Federal Agencies Could Improve Detection and Prevention (GAO-09-258)*

GAO Recommendation 3:

To help reduce the prevalence of seafood fraud and improve FDA's actions to detect and prevent seafood fraud, we are recommending that the Commissioner of the Food and Drug Administration update the Fish and Fisheries Products Hazards and Controls Guidance to reflect the seafood labeling requirements of the Food Allergen Labeling and Consumer Protection Act.

FDA Response:

The *Fish and Fishery Products Hazards and Controls Guidance* is a set of recommendations from FDA to the fish and fishery products processing industry regarding the hazards that are reasonably likely to occur in such products and providing suitable controls to minimize the risk of occurrence of those hazards. FDA intends to reflect the labeling requirements of the Food Allergen Labeling and Consumer Protection Act in the fourth edition of *Fish and Fishery Products Hazards and Controls Guidance*. The document is currently in Agency clearance.

GAO Recommendation 4:

To maximize the efficiency and effectiveness of each agency's efforts to detect and prevent seafood fraud and to increase interagency collaboration, improve information sharing, and reduce overlaps, we recommend that the Commissioner of Customs and Border Protection, the Under Secretary of Commerce for Oceans and Atmosphere, and Commissioner of the Food and Drug Administration develop goals, strategies, and mechanisms to share information and resources related to seafood fraud detection and prevention across agency boundaries.

FDA Response:

Three key federal agencies share responsibilities for detecting and preventing seafood fraud. According to GAO's report, each Agency mostly acts alone and has not implemented key practices to begin collaborating more effectively. FDA agrees that it should collaborate with the other federal agencies to maximize efficient use of resources and is currently collaborating with National Marine Fisheries Service (NMFS) to develop an inter-agency library of seafood species standards. As mentioned by GAO, FDA also has collaborated, leveraged resources, and shared data with the Department of Homeland Security's Customs and Border Protection (CBP) in efforts such as Operation Guardian and in sharing weekly refusal reports.

In December of 2007, Department of Homeland Security, Immigration, and Customs Enforcement (ICE) implemented Operation Guardian to combat the growth in the importation and distribution of a variety of consumer products, which potentially threaten the health and safety of U.S. consumers. Since the inception of Operation Guardian, several joint enforcement efforts have been initiated against imported food, including seafood, involving ICE, CBP, and FDA. In May 2007, Guardian regional working groups composed of ICE, CBP, and FDA personnel were formed throughout the country. Sixty-one agents from FDA's Office of Criminal Investigations (OCI) are assigned to the ICE-chaired 35 regional working groups.

FDA's General Comments to the U.S. Government Accountability Office's (GAO) Draft Report Entitled, *Seafood Fraud – FDA Program Changes and Better Collaboration Among Key Federal Agencies Could Improve Detection and Prevention (GAO-09-258)*

Guardian headquarters representatives and regional working group members generally meet monthly to evaluate ongoing investigations and incoming fraud allegations, and to initiate new enforcement actions. Several of these investigations are targeted at individuals and entities responsible for the fraudulent importation, through transshipment, of seafood subject to CBP anti-dumping duties and FDA import alerts which are intended to prevent seafood containing unapproved antibiotics from entering the United States. ICE and OCI will continue cooperative investigative efforts into such areas as transshipment or adulteration of seafood products. In addition, collaboration between CBP and the FDA has resulted in regulatory actions and import alerts.

Some other specific examples of FDA's collaborative efforts include the following:

- FDA collected a sample of "langostino" meats through the Port of Blaine, Washington that FDA determined were freshwater crawfish instead of langostinos. Freshwater crawfish from China are subject to an anti-dumping duty of 223 percent. An initial review of entries indicated that there were at least 23 entries of "langostino" meat valued at $2.3 million. FDA's report of these findings to members of the ICE resulted in the collection of $2.9 million in tariff duties from a Canadian firm that exports seafood to the United States. The firm admitted that it had misclassified tariff duties on freshwater crawfish from China.

- FDA participated in a joint investigation with United States Fish &Wildlife Service (USFWS) and ICE that involved the smuggling of illegal caviar and the sale of adulterated and misbranded product throughout the United States. Marky's Caviar, also doing business as Optimus, Inc., purchased sturgeon caviar smuggled into the United States by couriers inside suitcases without a CBP declaration or USFWS inspection. The investigation determined that the sturgeon caviar and other fish roe sold were adulterated and misbranded. The caviar was adulterated by mixing Beluga caviar with less valuable caviar. The caviar was misbranded because it was labeled as Russian caviar when it was in fact Chinese. On April 15, 2005, the firm was sentenced to pay a $1 million criminal fine into the Lacey Act Reward Fund, to adhere to a stringent wildlife compliance plan, and 5 years probation. The sentence was imposed for knowingly purchasing sturgeon caviar with false wildlife invoices, knowingly purchasing smuggled caviar, and failing to exercise due care in purchasing smuggled caviar.

GAO Recommendation 5:

To maximize the efficiency and effectiveness of each agency's efforts to detect and prevent seafood fraud and to increase interagency collaboration, improve information sharing, and reduce overlaps, we recommend that the Commissioner of Customs and Border Protection, the Under Secretary of Commerce for Oceans and Atmosphere, and Commissioner of the Food and Drug Administration create a federal agency-wide library of seafood species standards.

FDA's General Comments to the U.S. Government Accountability Office's (GAO) Draft Report Entitled, *Seafood Fraud – FDA Program Changes and Better Collaboration Among Key Federal Agencies Could Improve Detection and Prevention (GAO-09-258)*

FDA Response:

FDA is collaborating with other Federal agencies in developing a library of seafood species standards. FDA held a two-day workshop in January 2008 entitled *Regulatory Applications of DNA-Barcoding for the Species Identification of Fish* which was attended by members of the NMFS National Seafood Inspection Laboratory, the NMFS Northwest Fisheries Science Center, the Smithsonian National Museum of Natural History, the Canadian Center for DNA Barcoding, and the International Consortium for the Barcode of Life. In collaboration with the Smithsonian NMNH and the Canadian Center for DNA Barcoding, FDA published a paper in 2008 entitled *Potential Use of DNA Barcodes in Regulatory Science: Applications of the Regulatory Fish Encyclopedia* in the Journal of Food Protection. This paper laid the groundwork for DNA barcoding at FDA and provided a bridge for the new methodology with FDA's established set of reference tissues used in the old protein isoelectric focusing method. FDA also published an FDA Laboratory Information Bulletin method in 2008 entitled, *A Protocol for Validation of DNA-Barcoding for the Species Identification of Fish for FDA Regulatory Compliance*, which is currently in the final stages of a three laboratory inter-lab trial. This trial is the first step necessary for performing a full collaborative study required for international accreditation.

FDA has also established a contract with the Smithsonian National Museum of Natural History for authentication and vouchering of seafood standards collected by FDA. FDA believes that this contract could potentially be the foundation for the GAO-recommended library of seafood species standards, and will be discussing that possibility with NMFS and CBP.

APPENDIX IV. COMMENTS FROM THE DEPARTMENT OF HOMELAND SECURITY

U.S. Department of Homeland Security
Washington, DC 20528

Homeland Security

January 30, 2009

Ms. Lisa Shames
Director, Natural Resources and Environment
Government Accountability Office
441 G Street, NW
Washington, DC 20548

Dear Ms. Shames:

The Department of Homeland Security (DHS) appreciates the opportunity to review and provide comment on the Government Accountability Office's (GAO) draft report titled, *SEAFOOD FRAUD: FDA Program Changes and Better Collaboration among Key Federal Agencies Could Improve Detection and Prevention* (GAO-09-258).

DHS generally concurs with the report's two recommendations. Following are our recommendation-specific comments; technical comments were provided under separate cover.

Recommendation 1: Develop goals, strategies, and mechanisms to share information and resources related to seafood fraud detection and prevention across agency boundaries.

Response: Concur. Customs and Border Protection (CBP) will consider issues related to seafood fraud in its regular assessment of trade risk and its annual planning process in a continuing an ongoing effort to address these matters.

CBP was an active member of the President's interagency work group addressing import safety concerns. The interagency work group developed the Import Safety Action Plan, which acknowledges multiple areas of import safety concerns including information sharing among agencies. The action plan recognizes that agency collaboration, including the International Trade Data System (ITDS), is essential for addressing import safety concerns effectively. ITDS is a single window information integration program that will allow agencies to collaborate in the identification and prioritization of high-risk shipments, including those considered import safety risks and anti-dumping risks, such as those identified in this report.

On the heels of the President's Action Plan for Import Safety, the implementation of interagency workgroups led DHS agencies to conduct special operations and other efforts targeted at addressing health and safety issues. Throughout the report, GAO makes note of these special operations, specifically Operation Guardian, where interagency collaboration is demonstrated.

Recommendation 2: Create a federal agency-wide library of seafood species standards.

Response: Concur. CBP would not oppose the development of a national agency database provided that CBP be issued the funding and the authority to maintain the database.

CBP currently uses "Gen-Bank" which is a DNA Sequence Library of different species that is maintained by the National Institutes of Health/National Center for Biotechnical Information. This database accepts DNA Sequence Data from around the world and is truly international in nature and wide ranging. Prior to the entry of information into the database the information is reviewed by a panel of NIH DNA specialists. Being international in nature, it is more widely accepted around the world for a number of purposes. CBP frequently uses Gen-Bank and has not identified problems or issues with the data to date.

The proposal to create an agency wide database would appear to be duplicative of the Gen-Bank but only on a smaller National level. CBP would not oppose the development of a national agency database provided that CBP be issued the funding and the authority to maintain the database. However, CBP believes that it would be more efficient for the agencies to work with NIH in the maintenance of Gen-Bank and assist in whatever shortcoming might be associated with that data bank.

Again, we thank you for the opportunity to review and provide comments on this draft report and look forward to working with you on future homeland security issues.

Sincerely,

Jerald E. Levine
Director
Departmental GAO/OIG Liaison Office

End Notes

[1] GAO, *Results Oriented Government: Practices That Can Help Enhance and Sustain Collaboration among Federal Agencies,* GAO-06-15 (Washington, D.C.: Oct. 21, 2005). The other three practices discussed in our report on effective collaboration are (1) develop mechanisms to monitor, evaluate, and report on results; (2) reinforce agency accountability for collaborative efforts through agency plans and reports; and (3) reinforce individual accountability for collaborative efforts through performance management systems.

[2] See our most recent series, GAO, *High-Risk Series: An Update,* GAO-09-271 (Washington, D.C.: January 2009). Also see, GAO, *High-Risk Series: An Update,* GAO-07-310 (Washington, D.C.: January 2007).

[3] ICE is the largest investigative branch of the Department of Homeland Security. According to ICE, the agency assigns special agents to, among other things, investigate manufacturers and importers allegedly involved in the duty evasion schemes and uses agents stationed in foreign countries to conduct investigations.

[4] Import alerts are designed to ensure that products from processors covered by the alert are detained and refused entry into the United States until the importer can prove that the imported product is safe and complies with all applicable regulations, generally by providing FDA with the results of third-party laboratory analysis of the product.

[5] Generally, the Congress establishes normal customs duties and authorizes the executive branch to impose special customs duties, such as antidumping and countervailing duties.

[6] The 2008 Farm Bill made catfish subject to mandatory inspection by USDA. The new law requires that USDA continuously inspect domestic catfish, and that imported catfish meet equivalent standards before being imported into the United States. The Secretary of Agriculture in consultation with the Commissioner of the Food and Drug Administration has until the end of 2009 to issue regulations to implement the new law.

[7] NFI is a seafood industry advocacy organization. According to NFI, its mission is to educate the American public about seafood safety, sustainability, and nutrition. NFI's nearly 400 members range from the owners of small, family-owned fishing vessels to large, nationally traded corporations and include representatives of business, education, and government.

[8] The third-party audit of the member's processes is the normal practice NFI uses to hold its member accountable, but it reserves the right to dismiss the member from the Better Seafood Bureau without a third-party inspection or to take a different action, depending upon the type of violation.

[9] GAO-06-15.

[10] All imported goods are reviewed for national security purposes in a screening process that occurs prior to the importer filing for entry of the imported goods.

[11] The Department of Commerce's definition of true, dusted shrimp is that having a coating of rice or wheat flour constituting between 4 and 10 percent of the product's total weight after being dusted, but prior to the shrimp being frozen. In conjunction with subject matter experts, CBP developed a set of characteristics that a product is required to meet to be considered true, dusted shrimp.

[12] Proposal to Establish Procedures for the Safe Processing and Importing of Fish and Fishery Products, 59 Fed. Reg. 4142 (Jan. 28, 1994).

[13] GAO, *Food Labeling: FDA Needs to Better Leverage Resources, Improve Oversight, and Effectively Use Available Data to Help Consumers Select Healthy Foods,* GAO-08-597 (Washington, D.C.: Sept. 9, 2008).

[14] GAO-08-597.

[15] The Maine Lobster Promotion Council, established by Maine in 1991, markets and promotes the sale of Maine lobster in local, regional, national, and world markets year- round. The council comprises harvesters, dealers, and public members from each of Maine's three regions.

[16] U.S. Food and Drug Administration, Center for Food Safety and Applied Nutrition, *Fish and Fisheries Products Hazards and Controls Guidance: Third Edition* (June 2001). See the following Web address: http://www.cfsan.fda.gov/~comm/haccp4.html.

[17] GAO-06-15.

[18] NMFS inspected 63 foreign facilities from 2007 through November 2008, and FDA inspected 112 foreign facilities from October 2006 through June 2008.

[19] GAO, *Food Safety: FDA's Imported Seafood Safety Program Shows Some Progress, but Further Improvements Are Needed*, GAO-04-246 (Washington, D.C.: Jan. 30, 2004).

[20] Isoelectrophoresis is a method of testing that uses an electric field to separate proteins for a species to create a distinct pattern that can be used to compare with and identify a seafood species, among other types of species. This method of testing has been authenticated as an official method for species identification by an international accreditation agency.

[21] GenBank is an open-access DNA sequence database that is maintained at the National Institutes of Health's National Center for Biotechnology Information. GenBank receives DNA sequences from laboratories throughout the world. After being reviewed by National Center staff, these DNA sequences are placed in the public database.

[22] GAO, *Food Labeling: FDA Needs to Better Leverage Resources, Improve Oversight, and Effectively Use Available Data to Help Consumers Select Healthy Foods,* GAO-08-597 (Washington, D.C.: Sept. 9, 2008).

[23] GAO, *Results Oriented Government: Practices That Can Help Enhance and Sustain Collaboration among Federal Agencies,* GAO-06-15 (Washington, D.C.: Oct. 21, 2005).

In: Seafood Fraud and Safety: Background and... ISBN: 978-1-61761-500-9
Editors: Kaitlyn O. Harris

Chapter 3

SEAFOOD FORUM

David Sefcik

INTRODUCTION

On May 28, 2009 a Forum was held at the National Institute of Standards and Technology (NIST), in Gaithersburg Maryland (agenda attached) to discuss the problem and solutions to short net weight fraud occurring in the Seafood industry. The meeting was a collaborative effort attended by 30 participants representing Federal Agencies from Customs and Border Protection (CBP),Food and Drug Administration (FDA), National Marine Fisheries Service (NMFS), National Institute of Standards and Technology (NIST), and the U.S. Department of Agriculture (USDA); State Weights and Measures Officials from California, Florida, Maryland, Minnesota (teleconference), New York, Virginia, and Wisconsin; Trade Associations from American Frozen Food Institute (AFFI), Better Seafood Board (BSB), International Food Distributors Association (I FDA), National Association of Attorney Generals (NAAG), and National Fisheries Institute (NFI). Also participating were industry representatives from Maritime Products International, Pilot Brands, and Shinnecock Ocean Seafood - A Native American Company.

The purpose of the meeting was to achieve three primary objectives:

1. To clearly define the problem and economic impact from short net weight fraud.

2. To provide understanding of each agency's role, responsibility, authority, enforcement, and oversight of seafood and net content compliance.
3. To develop a plan for clear and immediate action and solutions to slow down and stop the problem of short weighing and to maintain the public's trust in the regulatory system.

BACKGROUND

In February, 2009, the U.S. Government Accountability Office (GAO) issued the report - *Seafood Fraud: FDA Program Changes and Better Collaboration among Key Federal Agencies Could Improve Detection and Prevention* (http://www.gao.gov/new.items/d09258.pdf). The study found deficiencies in how federal agencies (CBP, FDA, NMFS) enforce regulations for fraudulent practices in the seafood industry. One significant fraudulent practice mentioned in the report was short weighing.

Seafood industry leaders, through the National Fisheries Institute (NFI), have recognized that fraudulent practices such as trans-shipping fish from one country to another to avoid certain duties and tariffs, misidentification of species, and short weighing creates consumer mistrust in the industry as well as unfair trade practices that harm companies who follow the rules. NFI formed the Better Seafood Board (BSB) in 2007 to address these issues. The mission of the BSB is to highlight NFI members as companies that provide honest seafood products and to provide a mechanism for partners in the supply chain to report suspected fraudulent practices.

Industry efforts to fight fraud can only go so far. Government oversight and enforcement is essential to curtail these illegal and unethical practices. Due to limited enforcement of the existing fraud-related regulations, solutions are needed to protect the interests of the consumers and ethical businesses. With that in mind, the Weights and Measures Division of the National Institute of Standards and Technology (NIST), the National Conference on Weights and Measures (NCWM), and the Better Seafood Board (BSB) co-sponsored a one day forum to enhance collaboration between the various agencies and to develop clear solutions to the problem.

PROBLEM

The Better Seafood Board provided an overview of seafood fraud in the industry. This was followed by an industry panel, and concluded with the BSB providing an overview and summary of a February, 2009 report on Seafood Fraud.

Better Seafood Board

The Better Seafood Board (BSB) outlined current fraud practices. The presentation "Addressing Seafood Fraud" (see attached) demonstrated that fraud is not a new problem for the seafood industry. A 1988 report by GAO, an August 8, 2007 Congressional Research Service (CRS) Report for Congress and the February, 2009 GAO report highlighted fraud and deception in the Seafood industry. Misrepresentation (misbranding and adulteration) occur in three primary areas: short weighing, species substitution and trans-shipments to avoid duties. One goal of the BSB, created in 2007 in response to fraud, is to ensure integrity through the development of a process of accountability with its members. The BSB developed a call center for receiving and helping to resolve complaints. Several examples of short weighing offer letters and problems were presented.

Industry Panel

A panel of industry representatives consisting of Maritime International, Pilot Brands, and Shinnecock Ocean Seafood discussed the economic impact seafood fraud and short weighing has had on their company and industry. Below are some of the key points and views made by industry:

- The problem of intentional short weighing is global and not unique to the United States.
- The fraud is an enormously evasive problem because the industry is so fragmented in the number and volume of "players".
- Fraud is premeditated, organized and intentional in that product is being bought and sold at anywhere from 10% to 35% under the labeled net weight.

- Examples were presented to illustrate that it is difficult for honest companies to compete with competitors who are selling at a lower cost because of intentional short weighing. Those companies trying to do honest business are finding that the perceived "higher cost" of their goods (because they do not defraud) result in lower sales because of the difficulty to compete on the perceived cost basis. It is believed that some companies have resorted to fraudulent practices in order to survive.
- An industry panelist stated that this is clearly criminal activity and should be treated as such.
- It was recognized that short weighing fraud occurs primarily in the food service sector and in non-English speaking retail outlets, but was noted that fraud at retail and other outlets should not be ignored.
- It was mentioned that many end users of fraudulent packaged seafood are not even aware that they are being shorted. Consumers, food service businesses and distributors need to be educated on the definition of net weight and the types of fraud occurring.
- There is a perception that buyers of seafood commonly distrust the Seafood industry, particularly those in the food service and restaurant business.
- Wholesale outlets (e.g. Costco, Price Club and BJ's) are now selling seafood product in wholesale packages, which are introducing a new level of fraudulent packages into the marketplace. Packages at these new outlets should be inspected by Weights and Measures.

Better Seafood Board – GAO Report Summary

The Better Seafood Board provided a brief overview and summary of the February, 2009 GAO report on Seafood Fraud. The purpose of the GAO report was to review the actions Federal agencies take to detect and prevent seafood fraud, as well as the extent of agency collaboration. The types of fraud identified include trans-shipping to avoid customs duties, species substitution, over treating to increase water content, adding excessive ice to increase weight and short weighing. GAO found that Federal agencies (CBP, FDA, and NMFS) do not collaborate effectively to share information or coordinate resources. GAO recommended that guidance on seafood labeling for the 2004 Food Allergen Labeling and Consumer Protection Act (FALCPA) be updated, that Seafood Hazard Analysis and Critical Control Point (HACCP) regulations be amended to include control of economic fraud issues, and that goals, strategies and mechanisms for information and resource sharing among agencies be developed.

OVERSIGHT

Representatives from Food and Drug Administration (FDA), Customs and Border Protection (CBP), National Marine Fisheries Service (NMFS) and the State of Wisconsin Weights and Measures Division, discussed their roles, authority, and oversight of seafood. The regulatory roles are discussed below.

Food and Drug Administration

FDA reported that sampling programs are in place primarily to address food safety issues (including species substitution). This includes visual inspections of international imports at the border as well as inspections of plants domestically. FDA further reported that inspections are not done at retail or food service establishments. FDA's authority covers all adulterated and misbranded product. It was noted that a high priority is placed on food safety and limited resources make it difficult to expand into short weighing issues. FDA is working to gain a better understanding that will help address short weighing. FDA is discussing the possibility of developing a system that will identify and signal when economic fraud may be occurring. FDA asked for assistance from industry and state weights and measures programs to provide them as much detailed information as possible when problems occur. FDA will work to "get the word out" to industry that they are enforcing short weighing as a useful means of deterrence.

Customs and Border Protection

The CBP, which controls all U.S. ports of entry, recognized that short weighing has never really been a CBP priority. CBP focus is primarily on duties, transshipments of product and country of origin labeling. CBP stated that it assists other government agencies regarding the enforcement of their policies and regulations. The CBP appreciated being alerted to the problem of short weighing and will look into ways they can assist other government agencies regarding this problem. The CBP requested that they be contacted in the event known fraud is taking place. The CBP home page on their website was mentioned as one way to allow interested parties to submit (anonymously or otherwise) allegations of fraud.

National Marine Fisheries Service

NMFS reported that it provides a seafood inspection service program for industry. This is done at a fee and is voluntary; though to participate a company must agree to abide by the requirements of the program. NMFS inspects about 1/3 of all seafood consumed in the United States. NMFS noted that they are increasingly being asked to provide in-plant inspections in overseas facilities. They indicated that although this is currently cost prohibitive, the agency is considering ways of expanding its overseas presence. NMFS is committed to taking higher levels of enforcement action on economic fraud. They are reviewing ways to improve current standards, such as improving communications with FDA in notifying them of violations, and developing (follow up) programs to ensure companies are making changes and corrections to product found in violation. NMFS also acknowledged its lack of communication with CBP and the opportunity for improvement in this area.

State Weights and Measures

The Chief of Wisconsin's Weights and Measures Program discussed the use of NIST Handbook 133 "Checking the Net Content of Packaged Goods" as the basis of most states' test procedures for checking glazed and frozen block seafood net content. It was explained that each state decides whether to adopt and follow HB 133 test procedures into their laws and regulations. It was mentioned that the frequency of testing seafood varies by state, and that overall, it generally encompasses a small portion of all product inspections occurring in state weights and measures programs. In addition, it was noted that it is up to each state to determine enforcement action and this determination is often made on a case by case basis. The states represented at the meeting indicated that very little testing is done with glazed and frozen seafood because of the length of time it takes to perform a test, and because of the expense of the test due to the samples being destroyed. Weights and Measures inspectors primarily test at retail but will investigate complaints in food service establishments. To limit the time and cost, audit testing is usually done first as a screening tool to identify questionable product. The Wisconsin Weights and Measures Chief gave a presentation on the HB 133 test procedures on deglazing (see attached presentation "Ice Glazed Seafood: State Testing and Enforcement"). The state found that performing inspections more routinely has been effective and impactful at reducing short net weight problems.

Group Discussion Summary

The afternoon was spent in group discussion, exploring ways to better work together, identifying the obstacles and determining the support required to eliminate short weighing fraud. Five primary questions were explored:

1. Is there agreement that seafood fraud (short weighing) is a problem?
2. Is there a correlation between intentional economic fraud and food safety?
3. Does government have the tools and authority (legislative, regulatory, resources, appropriation) to be effective against seafood fraud?
4. What Seafood industry actions (e.g. education) are required?
5. What needs to occur to achieve effective collaboration between federal agencies?

Is there Agreement that Seafoodfraud (Short Weighing) is a Problem?

Forum participants were in agreement that economic fraud is occurring in the Seafood industry and that a concerted effort must be taken to address this practice. It was also agreed that intentional short weighing of 10% to 35% creates a significant economic impact to consumers and the government (e.g. tax revenue) and threatens regulatory integrity and equity in the market place.

Is there a Correlation between Intentional Economic fraud and food Safety?

Discussion ensued as to the correlation between intentional economic fraud and food safety. The argument was made that if a company willfully cheats customers and cuts corners in one area, that other areas such as food safety controls may also be overlooked. Discussion centered on the benefit of focusing on short net weight as a tool to identify companies with poor practices that could lead to food safety issues. Taking enforcement action on companies that practice short weighing will also have a positive impact on food safety. Federal agencies (FDA in particular) were challenged to consider this strategy and approach as part of their enforcement activities. NMFS will be looking into ways to assist smaller companies in providing net weight testing services at a discount.

Does Government Have the Tools and Authority (Legislative, Regulatory, Resources, Appropriation) to be Effective against Seafood Fraud?

Overall, the CBP, FDA, and NMFS believed they currently have the legislative authority to be effective. Discussion centered around the need for clarity on interpretations of current laws with examples being discussed such as; Can water that is added as glazing, and which is identified as part of the products description and as an ingredient, be included in the net weight? What constitutes a marinade? Should glazing standards (% and quality of glaze) be developed to promote ease of net content checking? One specific scenario discussed was "if a label states that the contents contains fish and water, and the content of fish is 32 kg and the glaze is 8 kg, can the package be sold as a 40 kg net weight package? FDA requested that the best approach to answering these questions is to pose specific questions in writing directly to the FDA. FDA will then review and provide an official response back to the industry or regulatory agency with clarity and guidance. The BSB will undertake a review of questions that need clarification and submit them to FDA.

Discussion continued on how to develop risk profiles of companies that may be violators. Use of prior violations was identified as one source, but it was noted that broad enforcement should be a collaborative Federal agency effort in order to be most effective. FDA “import alerts” were also identified as an effective tool to identify seafood fraud. NMFS discussed the justification and use of "market orders" as a tool. The group acknowledged that this is a difficult process to implement and that other options should first be explored.

Publicizing company violations was also seen as an effective deterrent to fraud. It was noted that highly publicized cases by media and FDA's re-issuing (see attached) of their 1991 letter regarding the fraudulent practice of including glaze as part of the weight of frozen seafood have had a very positive impact in changing behavior.

Another suggestion was to set up an industry compliance testing program for seafood product with requirements that, if met, would receive a "stamp of approval". The biodiesel industry's Fuel Quality System was used as an example.

There was consensus among Forum participants that the use of deterrents would be the most effective method due to limited resources of government agencies, the amount of product in the marketplace, and the fragmentation of the industry. Identifying the most effective deterrents would be one goal.

What Seafood Industry Actions (e. g. Education) are Required?

The group discussed ways the industry could benefit from education. An action such as the re-issuing of the 1991 FDA letter was seen as being an effective educational tool. The BSB agreed to take on this assignment in working with Federal, state and local agencies as well as with other trade associations in identifying effective methods of communication, types of information that should be communicated and the various sources with whom the information should be shared.

What Needs to Occur to Achieve Effective Collaboration between Federal Agencies?

Participants agreed that there needs to be a national seafood fraud workgroup formed to continue discussion and collaboration between Federal and state agencies. The goal of the workgroup would be to plan, coordinate and implement a collaborative investigative and enforcement effort between state and Federal agencies. NIST will follow-up with this request to coordinate a conference call or meeting.

SUMMARY OF ACTION ITEMS AND NEXT STEPS

- All participants will share knowledge gained from this Forum with their respective agency, company, or trade association in an effort to continue creating awareness of seafood fraud and in seeking solutions to the problem.
- States and industry will provide FDA with as much detailed information as possible when problems occur. Those present from FDA and CBP offered to be the contact persons.
- FDA will work to publicize to the industry that there will be greater enforcement activity in the area of short weighing. FDA will look into such areas as publicizing cases where seafood fraud is found, with the intent of being a deterrent to future activity of this kind. FDA will also review the correlation between short net weight cases and food safety to determine whether this should be considered part of FDA enforcement activities.
- CBP will look for ways to work closer with FDA regarding enforcement of policies and regulations that impact seafood fraud.
- NMFS will continue to look for opportunity for greater voluntary industry compliance specifically in its overseas inspections of plants. In addition, ways to assist small companies domestically will be reviewed. NMFS will also consider ways to improve current standard communications with FDA in notifying them of violations and developing (follow up) programs to ensure companies are making changes and corrective action to seafood products found in violation. NMFS enforcement office will become more involved with economic fraud.
- BSB will determine which regulatory guidelines and regulations need clarification and provide a list of these questions to the FDA for response. BSB will work with the Federal agencies, states and local agencies, trade associations and industry to identify effective educational tools.
- NIST will coordinate a conference call with FDA, CBP, NMFS and state weights and measures programs so that those involved with investigative and enforcement activities can coordinate their efforts in addressing seafood fraud.

In: Seafood Fraud and Safety: Background and... ISBN: 978-1-61761-500-9
Editors: Kaitlyn O. Harris

Chapter 4

SEAFOOD SAFETY: BACKGROUND AND ISSUES

Geoffrey S. Becker and Harold F. Upton

SUMMARY

Although seafood consumption can contribute to a healthy diet, some fish and shellfish can cause foodborne illnesses or contain environmental contaminants. Are current food safety programs sufficiently protecting consumers, and if not, what changes should be considered? A complicating factor is that most of the seafood consumed in the United States is from imports.

The Food and Drug Administration (FDA) within the Department of Health and Human Services plays the lead role in ensuring the safety of both domestic and imported fish and shellfish, but other agencies, including the National Marine Fisheries Service in the Department of Commerce and the Food Safety and Inspection Service in the U.S. Department of Agriculture, also have notable responsibilities.

In the 111th Congress, the food safety bills specific to seafood include S. 92, aimed at violative seafood imports, and H.R. 1370, authorizing $15 million annually to strengthen coordination between agencies on seafood safety and quality, particularly regarding imports. On July 29, 2009, the House approved H.R. 2749, a revised version of H.R. 759. H.R. 2749 takes a comprehensive approach to food safety, including seafood. On November 18, 2009, the Senate Health, Education, Labor, and Pensions Committee approved S. 510, another comprehensive approach to food safety.

Seafood Safety Risks

Studies and dietary recommendations have suggested that increased consumption of seafood can contribute to a more healthful diet.[1] Nonetheless, seafood consumption is not without risk. Although hazards caused by microbes and naturally occurring toxins in seafood have been well characterized, the public health burden has been difficult to quantify or to assess over time due to data limitations. Foodborne illness data are prone to under-reporting, and in many cases the cause of the illness (called the food vehicle) may not be determined. The U.S. Centers for Disease Control and Prevention (CDC) published data for approximately 3,550 reported outbreaks of foodborne illness that occurred during calendar years 2004 through 2006. A causative food vehicle was reported for about 1,900 of the outbreaks. Of these, seafood (finfish or shellfish) was reported as a vehicle in 310 (16% of the 1,900) outbreaks. In comparison, red meats were reported in 325 (17%) outbreaks, and poultry in about 290 (15%).[2] To put these data in context, annual U.S. per capita consumption of seafood was about 16 pounds in 2005, compared with 110 pounds for red meats and 74 pounds for poultry.[3] Therefore, when compared to the consumption levels of beef and poultry, seafood is responsible for a disproportionate number of outbreaks. However, an average outbreak related to seafood consumption generally affects a smaller number of people (i.e., fewer individual cases per outbreak).

Naturally occurring toxins were involved in more than half of all seafood-associated outbreaks with known or suspected causes in 2004-2006. Such toxins are primarily ciguatera, from certain tropical reef-dwelling finfish, and scombroid poisoning, which develops in some species after harvest due to inadequate refrigeration. Other common problems are norovirus, the bacterium *Vibrio parahaemolyticus* in raw shellfish, and various other pathogens, such as *Clostridium botulinum,* in processed seafood products.

The Institute of Medicine (IOM) has cited other risks of consuming seafood: that of high levels of chemical and heavy metal pollutants from the environment such as mercury, lead, polychlorinated biphenyls (PCBs), and pesticides. Some of these problems, such as high mercury levels, are more evident in carnivorous fish at the top of the food chain, such as shark, swordfish, and bluefin tuna.[4] But the IOM also has noted that it has been difficult to quantify the risks of some of these hazards due to incomplete data, the complexity of dietary and nondietary contaminant exposures, and the fact that certain health effects such as cancer develop over a much longer period than microbial illnesses.[5]

Worldwide, nearly one-half of all seafood consumed as food comes from aquaculture. Aquacultured (farm-raised) seafood also may contain potentially harmful chemicals. This was illustrated on June 28, 2007, when the Food and Drug Administration (FDA, in the U.S. Department of Health and Human Services) issued an import alert for the "Detention Without Physical Examination" of all aquacultured products of certain fish species from China. FDA said it issued the notice after targeted sampling in the prior year "repeatedly found that farm-raised seafood imported from China were contaminated with antimicrobial agents that are not approved for this use in the United States."[6]

Increased imports, including from many other Asian countries in addition to China, have complicated efforts to protect consumers from unsafe fish and shellfish. In 1995, imports already constituted more than half of U.S. per-capita seafood consumption; by 2008 they reached 83%.[7] By 2005, more than 150 countries were exporting seafood to the United States, FDA observed.

Current Inspection Programs

FDA Safety Inspection

FDA has primary responsibility for the safety of all domestic and imported foods, including seafood, under the Federal Food, Drug, and Cosmetic Act (FFDCA), as amended (21 U.S.C. Sec. 301 *et seq*.). Excepted are most meat and poultry products, which the U.S. Department of Agriculture's Food Safety and Inspection Service (F SIS) inspects under other statutory authorities. The FFDCA requires that all such foods be safe, wholesome, and accurately labeled. FDA-regulated foods may be deemed adulterated or misbranded for a variety of statutorily prescribed reasons. FDA also sets maximum safe levels of unavoidable toxic substances in foods, including fish, and requires that all domestic and foreign food manufacturing facilities adhere to Good Manufacturing Practices (GMPs; 21 C.F.R. Part 110), which address safe handling and plant sanitation. Generally exempt are establishments such as farms, including fish farms, that merely raise and/or harvest a raw commodity.

Seafood is one of the few FDA-regulated food groups further regulated under a system of risk prevention controls known as HACCP, for Hazard Analysis and Critical Control Points. Under HACCP, domestic processors are to prepare site- and product-specific plans that analyze potential safety hazards, find where they are likely to occur during processing, identify control points and how they will be monitored, and determine how hazards will be controlled. Importers of foreign seafood must take steps to verify that the products obtained from foreign processors are in compliance with the HACCP rules.[8]

As the 2001 GAO report (see footnote 3) noted, if a processor determines and FDA inspectors agree that a particular product is of low risk, no plan is needed; therefore not all firms necessarily will have a plan. Moreover, fishing vessels are exempt, unless they do more than minimal processing. Following publication of its HACCP rule, FDA sought to inspect all regulated seafood establishments to ensure that HACCP was being implemented, and to continue to visit each one annually. The agency recently reported that there are approximately 13,400 domestic seafood processing establishments, and that it had inspected a total of 3,066 in 2004, 2,830 in 2005, and 2,456 in 2006.[9]

The FFDCA empowers FDA to refuse entry to any food import if it "appears," based on a physical examination or otherwise, to be adulterated, misbranded, or in violation of the law. In exercising its oversight, the agency relies on a system of bonding and of prior notifications by importers and document reviews at points of entry (ports). From lists of these entries, the agency selects products for physical examination and/or testing to determine whether they contain adulterants. FDA inspected, or tested for contaminants, less than 2% of an estimated 860,000 seafood shipments in 2006.[10]

Foreign seafood processors are subject to the same requirements, including HACCP, as domestic firms, and the U.S. importers of their products must take "affirmative steps" to help ensure that these requirements are being met. FDA conducts inspections to check compliance of these importers and of selected foreign processors (81 processors inspected in 10 countries in 2004, 72 in 10 countries in FY2005, and 68 in 8 countries in 2006), focusing on those that are major exporters to the United States and on developing countries less likely to have sophisticated safeguards. FDA counted approximately 14,900 registered foreign exporters of seafood to the United States, but observed that "a great many more foreign firms are involved in the processing of the products that eventually are shipped to the U.S."[11]

FDA is exploring the potential use of third parties to certify foreign processors of aquacultured shrimp for compliance with FDA's seafood HACCP regulations. During Phase I of the pilot program, FDA screened requests from third-party certification bodies (private, nongovernmental, other federal government, and state government) to participate in the pilot; it announced six participants in late 2008.[12] During Phase II, FDA is conducting onsite audits of programs by accompanying inspectors during certification inspections. FDA said it intends to evaluate potential use of the pilot program for augmenting agency efforts, determining inspection priorities, and adjusting "may proceed" (i.e., entry without FDA staff review) rates for imports of aquacultured shrimp or other food products.[13]

Shellfish Safety

Under provisions of the FFDCA and the Public Health Service Act, FDA cooperates with 23 coastal shellfish-producing states and some foreign countries in a National Shellfish Sanitation Program (NSSP) to promote the safe production of fresh and frozen molluscan shellfish—oysters, clams, and mussels—for human consumption.[14] FDA works through the Interstate Shellfish Sanitation Conference (ISSC), an organization of state shellfish regulators who in turn adopt state and local laws, based on an NSSP "model ordinance," to ensure that shellfish in their jurisdictions are safe and sanitary. An objective of these laws is to ensure that products can be traced to harvest waters that are safe. Generally, dealers must be listed with their state regulatory agency in order to ship shellfish products commercially.[15]

On October 17, 2009, at the biennial meeting of the IS SC, FDA announced its intention to implement new requirements for post-harvest processing of Gulf of Mexico oysters to reduce the number of illnesses resulting from *Vibro vulnificus*.[16] *V. vulnificus* is a species of pathogenic bacteria found naturally in marine environments. Infections are more likely to occur during warm months, when bacteria counts are higher. *V. vulnificus* causes disease in individuals who eat contaminated oysters that are raw or undercooked. In healthy people, *V. vulnificus* can cause diarrhea, vomiting, and abdominal pain, but usually without long-term consequences. However, in immunocompromised individuals, *V. vulnificus* can invade the bloodstream and cause life- threatening symptoms. According to FDA, 30 severe cases occur each year with a 50% mortality rate. [17]

Since 2001, collaborative efforts of the ISSC to reduce illness associated with *V. vulnificus* have included disseminating educational materials and refrigerating oysters after harvests. However, according to FDA, the program goal of a 60% reduction in severe illnesses and death has not been attained. This prompted FDA to announce changes to its policy on raw shellfish that would be included in the update of *Fish and Fishery Products Hazards and Controls Guidance.*[18] FDA would require the application of post-harvest processing (PHP) technologies such as individual quick freezing with frozen storage, high hydrostatic pressure, mild heat, and low-dose gamma irradiation. PHP would be required by the summer of 2011 for oysters harvested during the months of April through October.

According to Gulf of Mexico seafood industry representatives, PHP infrastructure is not in place and the PHP requirements would be costly to implement and would threaten jobs.[19] The industry estimates that oyster prices could double or triple.[20] According to FDA, 40% of Gulf oysters are harvested during the warm months and about half are eaten raw.[21] The Gulf coast oyster industry already subjects about 15% of oysters to PHP methods, and the cost of PHP would be three to four cents per oyster.[22] Restaurant owners contend that PHP makes oysters less desirable to their customers because it changes the texture and flavor of the oyster. FDA counters that a study of oyster consumers has shown that most cannot recognize the difference between PHP-treated and raw oysters.

Several bills were introduced in response to this controversy, including identical House and Senate versions of the Gulf Oyster Protection Act of 2009 (S. 2735 and H.R. 4022) as well as the Gulf Oyster Industry Jobs Protection Act (S. 2752). All three bills would prohibit the use of federal funds to implement the FDA policy to require PHP. S. 2752 would also require the Secretary to conduct an education campaign to increase awareness of the risks associated with consuming raw oysters and to develop analyses for the appropriate committees of Congress when regulation or guidance is proposed that affects the harvesting, processing, or transportation of seafood harvested in the United States. On November 13, 2009, the FDA announced that it will conduct a study to assess how PHP or equivalent controls can be implemented.[23] FDA plans to confer with all parties to discuss their concerns while taking actions to assist the seafood industry.

NOAA Voluntary Inspection

Within the Department of Commerce, the National Marine Fisheries Service (NMFS) of the National Oceanic and Atmospheric Administration (NOAA) administers a voluntary seafood inspection program under authority of the Agricultural Marketing Act of 1946 (7 U.S.C. Sec. 1621 *et seq*.). The program offers additional levels and types of inspection that exceed the FDA HACCP-based requirements, which program participants also must meet. Examples include on-site NOAA inspections during production hours, certification that plants or vessels meet specified sanitation requirements, quality inspections of individual product lots, and/or laboratory testing of products, among other services. These services are provided on a fee-for-service basis and entitle participants to use various official grading and labeling marks, which are viewed as making their products more attractive to buyers. Exporters are often users of these services, in part because of foreign buyer requirements.

In 2006, NOAA reported active Seafood Inspection Program contracts with 377 firms, although the participant number changes constantly. The number of foreign participants (in 2008, more than 50) has increased recently due to such firms' desire to comply with requirements of the FDA import alert on aquacultured products from China. The number of participating firms is a small fraction of all seafood facilities, but they produce a significant portion of the total volume: in 2006, the NOAA voluntary program inspected 1.9 billion pounds or 33% of the total seafood consumed in the United States.[24]

In January 2008, the Government Accountability Office (GAO) issued a report on federal oversight of food safety.[25] The report included a recommendation for NOAA and FDA to develop an agreement to use NOAA's inspection program resources to carry out inspections on FDA's behalf and to augment FDA's program for imported seafood with inspection training and product sampling services. On October 26, 2009, NOAA and FDA issued a memorandum of understanding to reinforce the agencies' cooperative efforts. The agreement formalizes procedures for sharing information and for FDA to consider the results of NOAA inspections when determining the frequency of FDA seafood inspections.

GAO Report on Seafood Fraud

GAO in February 2009 reported that the three principal U.S. agencies that share responsibility for detecting and preventing seafood fraud—Customs and Border Protection (CBP) within the Department of Homeland Security, NMFS, and FDA—"do not effectively collaborate with each other. Specifically, they have not identified a common goal, established joint strategies, or agreed on roles and responsibilities." GAO observed that "CBP and NMFS conduct several activities to help detect and prevent seafood fraud, but FDA told GAO that it focuses on food safety and undertakes few fraud-related activities. Nonetheless, fraud can result in food safety problems. For example, fish that was mislabeled as a different species for financial gain has caused illnesses due to the presence of a potentially deadly toxin." The GAO report offered a number of recommendations to improve the detection and prevention of seafood fraud, including the addition of fraud detection provisions to the FDA seafood HACCP regulations.[26]

SELECTED LEGISLATION IN CONGRESS

Until the mid-1990s through the 104th Congress, many seafood safety proposals sought to put fish inspection on a par with that of meat and poultry. USDA's FSIS is required, under the Federal Meat Inspection Act (FMIA; 21 U.S.C. 601 *et seq.*) and the Poultry Products Inspection Act (PPIA; 21 U.S.C. 451 *et seq.*), to inspect all livestock and poultry both before and after they are slaughtered, and to be present whenever plants are processing meat and poultry products.[27]

Jurisdictional differences were among the reasons previous bills were not enacted. USDA and the House and Senate Agriculture Committees have long been responsible for meat and poultry inspection, while FDA, the Senate Committee on Health, Education, Labor, and Pensions (HELP), and the House Committee on Energy and Commerce have claimed jurisdiction over the safety of seafood (and other foods). Others such as the Senate Committee on Commerce, Science, and Transportation, also have had roles.

Seafood safety began to garner new attention in the 110th Congress, following a number of reports of contaminated foods, some from foreign sources, entering the food supply. Numerous congressional hearings and media reports in 2007 and 2008 brought wider public recognition of the role foreign producers now play in meeting U.S. demand for fish and shellfish. Underlining this awareness was the issuance of the Food and Water Watch report in May 2007, and the FDA action on Chinese seafood in June 2007.

2007 FDA Legislation

Wide-ranging FDA legislation (P.L. 110-85) adopted in 2007 includes a provision (§1006) requiring the report to Congress cited above in this CRS report. HHS also may enter into partnerships with states on inspection, under § 1006. Under § 1007, FDA must consult with NOAA on a report on environmental risks associated with genetically engineered seafood products, including the impact on wild fish stocks.

2008 Farm Bill

In June 2008, Congress enacted an omnibus farm bill (P.L. 110-246), with a section (§11016(b)) that newly designates "catfish," as defined by the Secretary of Agriculture, as an "amenable species"—that is, subject to mandatory inspection under the FMIA. The amendment will apply to companies that process catfish for food. It also directs USDA to "take into account the conditions under which the catfish is raised and transported to a processing establishment," and to consult with FDA. The inspection provision reportedly was urged by the U.S. catfish industry, which has faced strong competitive pressure from foreign catfish producers, particularly in Asia, where, U.S. interests allege, unacceptable types and levels of veterinary drugs are more frequently used. The conference report states the intent of Congress "that catfish be subject to continuous inspection and that imported catfish inspection programs be found to be equivalent under USDA regulations before foreign catfish may be imported into the United States." The report also noted that the Secretary already has authority under the FMIA to mandate inspection for other seafood species if he deems it appropriate. (He has not.)

FSIS is reportedly working on a proposed rule for implementing the catfish inspection regulations, but has not been able to meet the December 2009 deadline. In 2008, the agency reportedly estimated that it would have needed \$5.3 million and 91 staff in FY2009, and \$16.5 million and 95 staff in FY20 10, to implement the program. FSIS estimated that at least 22 processing plants, primarily in the southern United States, would have to be inspected, and at least 15 foreign countries would have to apply for an FSIS determination that their inspection systems are equivalent to the U.S. safety systems before any of their establishments could ship catfish to the United States.[28]

Also, § 11016(a) of the farm bill amends the Agricultural Marketing Act of 1946 (7 U.S.C. 1622) to require USDA to establish a voluntary grading program for catfish, which producers could opt into and pay for with user fees (as exists in other USDA quality grading programs authorized by the 1946 act). The section further authorizes producers of other farm-raised fish and shellfish species to apply for voluntary grading services.

Legislation in the 111th Congress

In the 111th Congress, several food safety bills have been introduced, and comprehensive legislation (H.R. 2749) has passed the House. The Senate also has a comprehensive bill (S. 510) that has been reported by the Senate Health, Education, Labor, and Pensions Committee. H.R. 2749 is a revised version of H.R. 759, both introduced by Representative Dingell. H.R. 2749 and S. 510 would focus primarily on FDA-regulated foods, including seafood, and would achieve their proposed reforms through the agency's existing structure and authorities. Both bills seek to increase the frequency of inspections, tighten record-keeping requirements, and mandate product recalls. Although H.R. 2749 and S. 510 would impose similar requirements, they would achieve them in different ways. Selected changes proposed by both bills include registration of all facilities; new requirements for record-keeping and for tracing products; mandatory FDA recall authority; authority for the agency to require imports to be certified as to their safety; and a requirement that all facilities have food safety plans that include hazard analyses and risk-based preventive controls.[29]

Another bill that would impact seafood producers and processors, particularly those participating in the NOAA-NMFS program, is H.R. 875 by Representative DeLauro. The DeLauro bill would create a new, independent Food Safety Administration within HHS but separate from FDA, to which all current food safety functions, personnel, and assets now at FDA would be transferred. Additionally, the DeLauro bill would move to the new agency all personnel and assets used to administer the NMFS seafood inspection program. At this time, however, it appears that H.R. 2749 has taken precedence as the primary House food safety bill.

As of late December 2009, two bills focusing specifically on seafood safety had been introduced. A measure (S. 92) by Senator Vitter would require the HHS Secretary to refuse all imports of seafood or seafood products from a country or exporter that does not meet requirements of the food and drug act (FFDCA) or is not likely to meet the requirements of any other federal food safety law. It also contains notification requirements for shipments that are refused entry. On the House side, H.R. 1370 by Representative Weiner would require the Secretaries of Commerce and HHS to enter into a memorandum of understanding for more cooperation and coordination on seafood safety activities (to include specific provisions listed in the bill), require an increase in laboratories certified to analyze seafood for compliance with federal law, and authorize $15 million in each of FY20 10 through FY20 14 to fund these and other activities outlined in the bill.

End Notes

[1] National Academy of Sciences (NAS), Institute of Medicine, Food and Nutrition Board, *Seafood Choices: Balancing Benefits and Risks*, 2007. The U.S. *Dietary Guidelines for Americans, 2005,* for example, cites limited evidence suggesting an association between consumption of fatty acids in fish and reduced risks of mortality from cardiovascular disease for the general population. Accessed at *http://www.health.gov/ dietaryguidelines/default.htm.*

[2] CRS analysis of CDC *Summary Statistics for Foodborne Outbreaks*, accessed January 2008 at http://www.cdc.gov/ foodborneoutbreaks/outbreak_data.htm. An outbreak is an incident involving at least two persons (cases).

[3] U.S. Department of Agriculture (USDA), Economic Research Service (ERS), *Food Availability (Per Capita) Data System*, accessed at *http://www.ers.usda.gov/Data/FoodConsumption/.* See also General Accounting Office (now Government Accountability Office, or GAO) report, *Food Safety: Federal Oversight of Seafood Does Not Sufficiently Protect Consumers* (GAO-01-204), January 2001.

[4] In March 2004, FDA and the Environmental Protection Agency, for example, issued a joint consumer advisory about mercury in fish and shellfish, directed at women who might become or are pregnant, nursing mothers, and young children. *Backgrounder for the 2004 FDA/EPA Consumer Advisory: What You Need to Know About Mercury in Fish and Shellfish*, accessed January 29, 2008, at *http://www.fda.gov/oc/opacom/ hottopics/mercury/backgrounder.html.*

[5] *Seafood Choices: Balancing Benefits and Risks*, 2007; and *Seafood Safety*, 1991, both National Academy of Sciences, Institute of Medicine, Food and Nutrition Board.

[6] FDA Import Alert #16-13, accessed January 23, 2008, at *http://www.fda.gov/ora//fiars/ ora_import_ia16131.html.* The fish species are catfish, basa (related to catfish), shrimp, dace (related to carp), and eel. The banned agents are nitrofuran, malachite green, and gentian violet, which have been found to be carcinogenic to laboratory animals; and fluoroquinolones, an antibiotic whose use may lead to antibiotic resistance. Under the import alert, FDA detains all covered products until the importer demonstrates, through

independent testing, that a representative sample of the product is free of these contaminants.

[7] U.S. Department of Commerce, National Oceanic and Atmospheric Administration, Fishwatch—U.S. *Seafood Facts,* at *http://www.nmfs.noaa.gov/fishwatch/trade_and_aquaculture.htm*

[8] Seafood HACCP regulations (at 7 U.S.C. Part 123) were published in final form in the December 18, 1995 *Federal Register* and became effective on December 18, 1997.

[9] FDA, Center for Food Safety and Applied Nutrition. Report to Congress, Food and Drug Administration Amendments Act of 2006, P.L. 110-85, Section 1006—Enhanced Aquaculture and Seafood Inspection, December 20, 2008. Also as required by Section 1006, the document discusses the feasibility of traceability systems for all catfish and seafood products in order to identify their processing plant of origin, and assesses the risks associated with particular contaminants and banned substances. Accessed March 31, 2009, at http://www.cfsan.fda.gov/~lrd/seartc08.html.

[10] Food and Water Watch, *Import Alert: Government Fails Consumers, Falls Short on Seafood Inspections*, May 2007.

[11] *Report to Congress on P.L. 110-85, Section 1006.* The report also noted that FDA conducted inspections of 657 seafood importers in 2004, 500 in 2005, and 529 in 2006, out of a total of 2,660.

[12] "FDA Announces Participants of Pilot Program for Third-Party Certification of Imported Aquacultured Shrimp," December 2, 2008 press release, accessed March 31, 2009, at http://www.fda.gov/bbs/topics/NEWS/2008/ NEW01919.html.

[13] 73 *Fed. Reg.* 39705-39708 (July 10, 2008). A separate FDA document issued in January 2009 sets forth the agency's criteria for others' use of *voluntary* third-party certification for food and animal feeds. More information on, and a link to, the FDA *Guidance for Industry: Voluntary Third-Party Certification Programs for Foods and Feeds* can be found in CRS Report R40443, *Food Safety: Selected Issues and Bills in the 111th Congress.*

[14] FDA is authorized to accept assistance from state and local authorities and others in the enforcement of its laws to assure food safety and to prevent the spread of communicable diseases, at 21 U.S.C. 372 in the FFDCA, and 42 U.S.C. 243 in the Public Health Service Act, respectively.

[15] See FDA, National Shellfish Sanitation Program, *Guide for the Control of Molluscan Shellfish, 2005*, at http://www.cfsan.fda.gov/~ear/nss3-toc.html.

[16] Michael R. Taylor, "Remarks to the ISSC Biennial Meeting," Manchester, NH, October 17, 2009, http://www.fda.gov/NewsEvents/Speeches/ucm187012.htm.

[17] FDA, "Questions and Answers about the Impact of FDA's New Policy to Prevent *Vibrio vulnificus* in Gulf Coast Raw, Untreated Oysters," press release, November 3, 2009, http://www.fda.gov/NewsEvents/Speeches/ ucm188800.htm. Hereinafter cited as FDA Questions and Answers 2009.

[18] The fourth edition is currently under development at FDA. The third edition is available at http://www.fda.gov/Food/FoodSafety/Product-SpecificInformation/Seafood/ SeafoodHACCP/default.htm

[19] Chris Kirkham, "Louisiana blasts new FDA rule requiring oysters to be sterilized to prevent rare bacterial illness," *The Times-Picayune*, October 28, 2009, *http://www.nola.com/dining/index.ssf/2009/10/*louisiana_blasts_fdajlan_to_l.html, Hereinafter cited as Kirkham 2009.

[20] Kirkham 2009.

[21] FDA, "Questions and Answers," 2009.

[22] FDA, "Questions and Answers," 2009.

[23] FDA, "FDA Statement on *Vibrio vulnificus* in Raw Oysters," press release, November 13, 2009, http://www.fda.gov/ NewsEvents/Newsroom/PressAnnouncements/ucm1905 13.htm.

[24] January 25, 2008, personal communication, Seafood Inspection Program, NOAA. Also see NOAA, USDC Seafood Inspection Program, at http://seafood

[25] U.S. Government Accountability Office, Federal Oversight of Food Safety: FDA's Food Protection Plan Proposes Positive First Steps, but Capacity to Carry Them Out Is Critical, GOA-08-435T, January 29, 2008, http://www.gao.gov/ new.items/d08435t.pdf.

[26] *Seafood Fraud: FDA Program Changes and Better Collaboration among Key Federal Agencies Could Improve Detection and Prevention,* GAO-09-258, February 19, 2009. The quotations are from the GAO highlights page, accessed March 31, 2009, at http://www.gao.gov/highlights/d09258high.pdf.

[27] See CRS Report RL32922, *Meat and Poultry Inspection: Background and Selected Issues*.

[28] "Details of USDA catfish inspection plan revealed in 5-page memo," *Food Chemical News*, December 1, 2008.

[29] The bill states that the new provisions for hazard analysis and risk-based controls, for example, are not to limit the HHS Secretary's authority to revise, issue, or enforce product-specific regulations already in force such as those for seafood HACCP. For more details on H.R. 2749 and S. 510, see CRS Report R40443, *Food Safety: Selected Issues and Bills in the 111th Congress*.

In: Seafood Fraud and Safety: Background and... ISBN: 978-1-61761-500-9
Editors: Kaitlyn O. Harris

Chapter 5

How FDA Regulates Seafood

FDA Consumer Health Information

FDA Detains Imports of Farm-Raised Chinese Seafood

On June 28, 2007, FDA announced a broader import control of farm-raised catfish, basa, shrimp, dace (related to carp), and eel from China. FDA will start to detain these products at the border until the shipments are proven to be free of residues from drugs that are not approved in the United States for use in farm-raised aquatic animals. The agency took this action to protect American consumers from unsafe residues detected in these products. There have been no reports of illnesses to date.

FDA is taking this strong step now because of continuing evidence that certain Chinese aquaculture products imported into the United States contain illegal substances. Aquaculture, also known as fish farming, involves raising fish in enclosed areas to be sold for food. Almost half of all imported seafood is from aquaculture, according to the U.S. Department of Commerce.

During targeted sampling, from October 2006 through May 2007, FDA repeatedly found that farm- raised seafood from China was contaminated with antimicrobial agents that are not approved for use in the United States. More specifically, the antimicrobials nitrofuran, malachite green, gentian violet, and flouroquinolones, were detected. Nitrofurans, malachite green, and gentian violet have been shown to cause cancer with long-term exposure in lab animals. The use of fluoroquinolones in food animals may increase antibiotic resistance, making it harder for this class of drugs to fight certain infections in people.

"Consumers should know that this is not an immediate public health hazard," says Robert Brackett, PhD, director of FDA's Center for Food Safety and Applied Nutrition. "The levels of contaminants that have been found are very low, and FDA is not advising consumers to destroy or return farm-raised seafood that they may have already purchased and have in their homes. The agency also is not seeking a recall of products already in the marketplace."

FDA is taking this action as a precautionary measure to prevent problems that may occur from longterm exposure to harmful residues. The agency is also concerned about the possible development of antibiotic resistance. "This action serves to keep contaminated products from entering the country so that they don't reach American consumers," Brackett says.

FDA, in collaboration with state regulatory counterparts, conducts in-plant inspections that focus on product safety, plant/food hygiene, economic fraud, and other compliance concerns.

Here's a look at how FDA works to protect consumers from unsafe seafood.

How do drug residues end up in fish? Some fish are given drugs to treat bacterial and parasitic diseases that cause major mortalities in fish. FDA's Center for Veterinary Medicine (CVM) regulates drugs given to animals. CVM conducts research to improve the drug approval process and expand the number of safe drugs available for fish production. CVM also develops methods to detect unapproved chemicals in fish tissues so that harmful drug residues don't wind up in the fish on your plate.

Is imported seafood required to meet the same standards as domestic seafood? Yes. Imported foods must be pure, wholesome, safe to eat, and produced under sanitary conditions. FDA requires imported seafood to be free of harmful residues. Importers must comply with regulations under the Federal Food, Drug and Cosmetic Act and the Fair Packaging and Labeling Act. In addition, seafood must be processed in accordance with FDA's Hazard Analysis and Critical Control Point (HACCP) regulations. A 1997 regulation, "Procedures for the Safe and Sanitary Processing and Importing of Fish and Fishery Products," requires seafood processors to identify food safety hazards and apply preventive measures to control hazards that could cause foodborne illness.

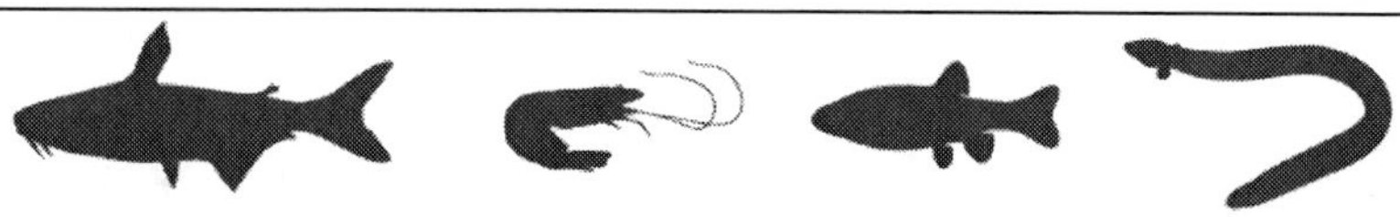

After FDA repeatedly found that farm-raised seafood from China was contaminated, the agency announced on June 28, 2007, a broader import control of all farm-raised catfish, basa, shrimp, dace(related to carp), and eel from China.

What other specific FDA regulatory programs focus on seafood?

- **National Shellfish Sanitation Program:** Administered by FDA, this program provides for the sanitary harvest and production of fresh and frozen molluscan shellfish (oysters, clams, and mussels). FDA conducts reviews of foreign and domestic molluscan shellfish safety programs.

- **Salmon Control Plan:** This is a voluntary, cooperative program among industry, FDA, and the Grocery Manufacturers Association/Food Products Association. It's designed to provide control over processing and plant sanitation, and to address concerns in the salmon canning industry.
- **Low-Acid Canned Food (LACF) Program:** To ensure safety from harmful bacteria or their toxins, especially the deadly Clostridium botulinum (C botulinum), in canned foods, regulations were established to ensure that commercial canning establishments apply proper processing, controls, such as heating the canned food at the proper temperature for a sufficient time to destroy the toxin-forming bacteria. Products such as canned tuna and salmon are examples of LACF seafood products.

How does FDA know when there is a safety concern associated with seafood?

FDA, in collaboration with state regulatory counterparts, conducts in-plant inspections that focus on product safety, plant/food hygiene, economic fraud, and other compliance concerns. FDA also receives notice of every seafood entry coming from a foreign country and selects entries from which to collect and analyze samples. FDA laboratories analyze samples for the presence of various safety hazards and contaminants, such as pathogens, chemical contaminants, unapproved food additives and drugs, pesticides, and toxins. Through close collaboration with the Centers for Disease Control (CDC) and state and foreign regulatory partners, FDA also learns of seafood safety concerns that arise through reports of illness potentially associated with seafood products.

An import alert identifies products that are suspected of violating the law so that FDA field personnel and U.S. Customs and Border Protection staff can stop these entries at the border prior to distribution in the United States.

What steps does FDA take when problems with seafood are detected?

For imported seafood, FDA has the authority to detain the food at the border to keep it from entering the country. This happens when FDA's analysis of such products indicate that they are not in compliance with the laws and regulations enforced by FDA. FDA can subsequently refuse entries of detained products if evidence of compliance is not provided by the importer or the importer does not correct the problem.

FDA has developed a number of import alerts that address problems found in seafood products in the past. An import alert identifies products that are suspected of violating the law so that FDA field personnel and U.S. Customs and Border Protection staff can stop these entries at the border prior to distribution in the United States. Usually, these import alerts will describe the products or firms that are subject to detention without physical examination. When products are detained without physical examination, the burden for demonstrating compliance of the product falls on the importer. Such compliance must be demonstrated before the product can enter U.S. commerce.

FDA can recommend criminal prosecution or injunction of responsible domestic firms and individuals, as well as seizure of contaminated products in commercial distribution within the U.S. FDA also works with domestic seafood processors to initiate voluntary recalls of contaminated products that may pose a safety concern to consumers.

What kind of research on seafood safety does FDA do?

FDA conducts research to better understand the nature and severity posed by various safety hazards, and other defects which may affect quality and economic integrity and to develop methods to minimize these risks. There are FDA laboratories specializing in seafood research on the Atlantic, Gulf, and Pacific coasts to address regional problems related to toxins and contaminants. FDA also has a facility in Laurel, Md., for conducting state-of-the-art research on drugs used in aquaculture.

What is the consumer's role in seafood safety?

As with any food, consumers should take precautions to reduce the risk of foodborne illness associated with sea-food. This includes properly selecting, preparing, and storing seafood. For example, consumers should only buy food from reputable sources and buy fresh seafood that is refrigerated or properly iced. Also, most seafood should be cooked to an internal temperature of 145°F. Some people are at greater risk for foodborne illness and should not eat raw or partially cooked fish or shellfish. This includes pregnant women, young children, older adults, and people with compromised immune systems.

For More Information

Questions and Answers on FDA's
Import Alert on Farm-Raised
Seafood from China

www.cfsan.fda.gov/~frf/seadwpe.html

Fresh and Frozen Seafood:
Serving It Safely
www.cfsan.fda.gov/~lrd/seafsafe.html

Food Safety for Moms-to-Be
www.cfsan.fda.gov/~pregnant/safemea. html
Timeline: Farm-Raised Fish
Imported from China
November 16, 2001 - June 28, 2007
www.fda.gov/consumer/updates/fishtimeline062807.html

In: Seafood Fraud and Safety: Background and... ISBN: 978-1-61761-500-9
Editors: Kaitlyn O. Harris

Chapter 6

TESTIMONY OF DON KRAEMER, DEPUTY DIRECTOR, OFFICE OF FOOD SAFETY, CENTER FOR FOOD SAFETY AND APPLIED NUTRITION, U.S. FOOD AND DRUG ADMINISTRATION, BEFORE THE U.S.-CHINA ECONOMIC AND SECURITY REVIEW COMMISSION, HEARING ON "CHINESE SEAFOOD: SAFETY AND TRADE ISSUES"

Members of the Commission, thank you for the opportunity to participate in today's hearing entitled *Chinese Seafood Imports: Safety and Trade Issues*. I am pleased to provide testimony for the U.S. Food and Drug Administration (FDA or the Agency) addressing one of your key issues for this hearing, which is to assess the health impact of imported Chinese seafood.

FDA'S SEAFOOD SAFETY PROGRAM

FDA has statutory authority and responsibility for the safety of all food, except for most meats, poultry and processed egg products, which are regulated by the U.S. Department of Agriculture.

With respect to seafood, FDA operates a mandatory safety program for all fish and fishery products under the provisions of the Federal Food, Drug and Cosmetic (FD&C) Act, the Public Health Service Act, and related regulations. The FDA program includes research, inspection, compliance, enforcement, outreach, and the development of regulations and guidance. As a cornerstone of that program, FDA publishes the *Fish and Fisheries Products Hazards and Controls Guidance*, an extensive compilation of the most up-to-date science and policy on the hazards that affect fish and fishery products and effective controls to prevent their occurrence. FDA is finalizing the fourth edition of this guidance document, which has become the foundation of fish and fishery product regulatory programs around the world.

FDA's program for fish and fishery products is comprehensive in its nature and is fully integrated into the food safety structures of FDA's Center for Food Safety and Applied Nutrition (CFSAN) and FDA's field organization, the Office of Regulatory Affairs (ORA). Because of the cold-blooded nature of fish and the nature of the aquatic environment in which they live, fish and fishery products pose unique food safety challenges which are quite different than those posed by land animals. FDA has developed extensive expertise in these areas over decades of regulating this commodity.

CFSAN experts are responsible for evaluating the hazard to public health presented by chemical and microbiological contaminants in fish and fishery products. FDA operates the Gulf Coast Seafood Laboratory in Alabama, which specializes in seafood microbiological, chemical and toxins research. In addition, seafood research is conducted at CFSAN's research laboratory in College Park, Maryland. FDA, in collaboration with the National Oceanic and Atmospheric Administration (NOAA) at the Department of Commerce, also represents the United States at the Codex Alimentarius Commission's Committee on Fish and Fishery Products, the international food safety standard setting body for this commodity.

INSPECTIONS

FDA's field staff is responsible for ensuring regulatory compliance for fish and fishery products produced in the United States and for those products imported from abroad. The field staff conducts inspections of fish and fishery product processing establishments, conducts follow-up investigations to track food-borne illnesses, and performs other activities designed to ensure the safety of these products. In FY 2007, FDA staff and state contractors conducted approximately 3,600 inspections of foreign and domestic seafood manufacturers, processors, importers, and storage facilities.

Processors of fish and fishery products are subject to FDA's Procedures for the Safe and Sanitary Processing and Importing of Fish and Fishery Products, commonly known as the Seafood Hazard Analysis Critical Control Point (HACCP) Regulation, in 21 C.F.R. Part 123. In short, this regulation requires both domestic and foreign processors of fish and fishery products to understand the food safety hazards associated with their process and product and, through a system of preventive controls, to keep those hazards from occurring.

The HACCP inspection approach is used by FDA during domestic and foreign inspections of seafood processors to focus its attention on the parts of seafood production and processing that are most likely to affect the safety of the product. In contrast to historical methods of evaluating processing practices on the day of the inspection, the HACCP approach allows FDA to evaluate processors' overall implementation of their HACCP systems over a period of time by having access to the firms' HACCP Plans, including monitoring, corrective action, and verification records. In this model, it is the seafood industry's responsibility to develop and implement HACCP controls and the regulatory agency's to ensure that the industry complies.

Every three years, FDA issues compliance programs that outline the Agency's field staff's inspection responsibilities. The Domestic Fish and Fisheries Products Compliance Program (CP 7303.842) and the Import Seafood Compliance Program (CP 7303.844) provide a priority list for inspection coverage based mostly on risk. Examples of high priority products include ready-to-eat products, such as hot or cold smoked fish, scombrotoxin-forming fish, such as tuna or mahi-mahi, aquacultured seafood products, and fish packed in reduced oxygen packages.

Annually, FDA determines a "work plan" for each FDA district office that outlines the Agency's field staff's domestic inspection responsibilities. This work plan focuses on areas that are a priority for the Agency, and allocates available resources.

Even though inspectional coverage is based primarily on product risk, FDA district offices may adjust that coverage to a particular establishment, such as one that may have been associated with a consumer complaint or illness or one with a poor compliance history. For example, the work plan may dictate that a processor be inspected annually, but if during an inspection the processor is found out of compliance re-inspection will occur more rapidly.

The regulatory sanctions that FDA has available to apply to domestic processors of fish and fishery products that are non-compliant are warning letters, seizure of products, injunction against further noncompliant practices, or prosecution of an individual or establishment.

Regulation of Food Imports

FDA's primary authority over imported food and other products under our jurisdiction derives from section 801 of the FD&C Act, which provides a broad statutory framework to ensure that the products are safe, wholesome, and accurately labeled.

When an FDA-regulated product is offered for import into U.S. commerce, Customs and Border Protection (CBP) procedures ensure that FDA is notified. If the product appears to be adulterated or misbranded, based on examination or other information such as prior history of the product, manufacturer or country, FDA will give notice advising the owner or consignee of the appearance of a violation and the right to provide evidence (such as a laboratory analysis by an independent laboratory) to rebut the appearance of the violation. In some circumstances, importers may request permission to recondition the product to bring it into compliance with applicable requirements and regulations. If the product is ultimately refused admission, it must be destroyed within 90 days unless reexported by the owner or consignee.

To better manage the increasing volume of imported products that we regulate, FDA currently screens electronically-submitted information on all incoming shipments, and then uses a risk-based approach to target our inspectional resources at products having the greatest potential for causing harm to public health. It is important to note that while FDA is not able to physically inspect a large percentage of import entries, all import entries are electronically screened for a variety of risk factors, using the Operational and Administrative System for Import Support (OASIS). OASIS is an automated system for processing and helping FDA make admissibility determinations for regulated products offered for import.

In 2002, Congress gave FDA significant new authorities to enhance protection of the food supply in the Public Health Security and Bioterrorism Preparedness and Response Act (the Bioterrorism Act). One of the most important provisions is the requirement that FDA be provided prior notice of food (including animal feed) that is imported or offered for import into the U.S. This advance information enables FDA, working closely with CBP, to more effectively target food that may be intentionally contaminated with a biological or chemical agent or which may pose a significant health risk to the American public. Suspect shipments then can be intercepted before they arrive in the U.S. and held for further examination. Prior notice can be submitted either through CBP's Automated Broker Interface/Automated Commercial System (ABI/ACS) or FDA's Prior Notice System Interface (PNSI). Currently, FDA receives approximately 33,400 prior notice submissions per business day.

FDA has numerous other tools and authorities that enable the Agency to take appropriate action regarding imported products. The Agency conducts some inspections of food manufacturers overseas. FDA also performs routine surveillance examinations of imported goods to check for compliance with U.S. requirements. Because of the large volume of FDA-regulated foods being exported from a large number of countries, it is not feasible to routinely inspect every shipment of foreign-produced foods at the point of import. We do, however, work with foreign governments and food producers to help ensure that imported food is produced, processed, and packed in accordance with U.S. requirements.

Another key tool for screening imported goods is the Import Alert. Import Alerts inform FDA field personnel that the Agency has sufficient evidence or other information about a particular product, producer, shipper or importer to believe the product does not meet U.S. requirements or is otherwise unsafe. On the basis of that evidence, FDA field personnel may detain the article that is being offered for entry into the U.S. without physically examining the product. When an Import Alert is issued and FDA detains a shipment, the owner or consignee has an opportunity to introduce evidence to demonstrate that the product is not violative.

FDA also performs laboratory analysis on a sampling of products offered for import into the U.S. and performs periodic filer evaluations to ensure that import data being provided to FDA is accurate. Certain violations relating to imported food may lead to civil or criminal charges.

Seafood Imports

More than 80 percent of all seafood consumed by Americans is imported. FDA regulates imported seafood products, including those from aquaculture, by conducting foreign manufacturer inspections, inspecting importers, and performing examinations and collecting surveillance samples of imported goods at the time of entry. FDA prioritizes these import-related activities based on the products' risk. High priority products, and foreign processors or importers of high priority products, are assigned higher priority for surveillance activities. Lower priority products, processors, and importers are sampled or inspected less frequently with remaining resources. Examples of high priority imported products are the same as those produced domestically and include ready-to-eat products such as hot or cold smoked fish, scombrotoxin-forming fish such as tuna or mahi-mahi, aquacultured seafood products, and fish packed in reduced oxygen packages.

It is the importer's responsibility to offer for entry into the United States product that is fully compliant with all applicable U.S. laws. As previously stated, under the Seafood HACCP Regulation, HACCP controls are required for both domestic and foreign processors of fish and fishery products. Additionally, the regulation requires that U.S. importers take certain steps to verify that their foreign suppliers meet the requirements of the regulation.

In fiscal year (FY) 2007, FDA processed approximately 868,000 entries of imported seafood, while our field staff performed more than 14,000 physical examinations of seafood imports and collected over 6,000 samples of domestic and imported seafood for analysis at FDA field laboratories.

Chinese Imports

China is a major producer, exporter, and importer of FDA-regulated products and it presents a diverse range of issues for the Agency. China is presently one of the world's largest producers and consumers of agricultural products, and a major supplier to the U.S. of seafood, canned vegetables, fruit juices, honey, and other processed foods. In the past, FDA has encountered compliance problems with several Chinese food exports, including lead and cadmium in ceramicware used to store and ship food, and staphylococcal contamination of canned mushrooms. While improvements have been made in these products, the safety of food and other products from China remains a concern for FDA, Congress, and American consumers. While these concerns are not unique to China, recent incidents have focused greater attention Chinese products.

Aquacultured Seafood

Aquacultured seafood is a fast-growing sector of the world food economy, accounting for approximately half of all seafood production worldwide. More than 80 percent of the seafood consumed in the U.S. is imported from approximately 130 countries, and over 40 percent of that seafood comes from aquaculture operations. By volume, China is the largest exporter of seafood to the U.S., and the second largest in terms of monetary value. In particular, China exports significant amounts of shrimp and catfish products, which represent two of the ten most consumed seafood products in the U.S.

As the aquaculture industry continues to grow, concern about the use of unapproved drugs and unsafe chemicals in aquaculture operations has increased significantly. There is clear scientific evidence that the use of unapproved antibiotics and other drugs and chemicals, such as malachite green, nitrofurans, fluoroquinolones, and gentian violet, can result in the presence of residues in the edible portions of aquacultured seafood. Fluoroquinolones are not approved for use in food fish and have been prohibited from extra-label use in the U.S. and many other parts of the world because of public health concern about the development of antimicrobial resistance. Moreover, prolonged exposure to nitrofurans, malachite green, and gentian violet, or their metabolites, has been shown to induce cancer in humans or animals. From a regulatory perspective, FDA has not approved any of these substances for use as drugs in aquacultured fish, nor are they generally recognized as safe or approved as food additives under the FD&C Act.

Since November 2001, FDA has tested shipments of aquacultured seafood products from China and other countries, and when residues of unapproved drugs have been found, has placed individual firms on Import Alert. In 2006, we significantly broadened these restrictions by issuing an Import Alert providing for the detention without physical examination of eel from anywhere in China due to findings of malachite green. In the course of an increased sampling program of imported Chinese aquacultured seafood which ran from October 1, 2006, through May 31, 2007, FDA continued to find residue of unapproved drugs in fish species including catfish, basa, shrimp, dace and eel.

Because the problems were seen in product from many different companies located in various parts of China, FDA imposed a countrywide Import Alert (IA #16-131) on all farm-raised catfish, basa, shrimp, dace and eel from China.

Shipments of products covered by Import Alert 16-131, which went into effect on June 28, 2007, are subject to detention without examination at the time they are offered for import into U.S. commerce. The shipments can be released by FDA after evidence is provided to overcome the appearance that the products are violative. Such evidence could include appropriate samples of the product properly analyzed for the aquaculture drugs of concern by a third party laboratory and found to be free of drug residues.

A producer that provides evidence that it has the appropriate controls and processes in place to ensure the safety of its product will be considered for removal from detention without physical examination (DWPE). The import alert recommends that the producer provide information in three areas, as follows:

1. Documentation showing that a minimum of five consecutive entries have been released by FDA based on third-party laboratory analysis verifying that the products do not contain the specified substances;
2. Documentation from an appropriate third party demonstrating that an inspection of the processor was conducted and that the seafood was processed in accordance with FDA's Seafood HACCP regulations, including controls for aquaculture drugs; and
3. Documentation that the processor is in compliance with all Chinese government requirements for exporting aquacultured seafood to the United States.

Since imposition of the countrywide Import Alert, FDA has detained 2,964 shipments of aquacultured seafood from China, and through laboratory testing, 1,387 of those shipments have been released into U.S. commerce.

Zhanjiang Guolian Aquatic Products Corporation, Zhanjiang, China (Guolian), one of the largest aquaculture shrimp exporting firms in that country, was the first Chinese company to request removal from DWPE under the Import Alert. The Chinese government's regulatory agency responsible for the export of food, the General Administration of Quality Supervision, Inspection and Quarantine (AQSIQ) certified Guolian's compliance with FDA's Seafood HACCP Regulation with respect to the control of unapproved aquaculture drugs. AQSIQ has expressed its intention to FDA to use Guolian as a model to show other aquaculture companies how to implement preventive controls for aquaculture drugs that will minimize the risk of unapproved drug residues in the product, facilitating their removal from the Import Alert.

FDA's consideration of Guolian's request for removal from DWPE consisted of the following actions.

1. FDA personnel reviewed documentation of the firm's HACCP controls for unapproved aquaculture drugs and performed an on-site audit of AQSIQ's physical inspection of the firm's processing plant in China, which included a review of the documentation for aquaculture drug process controls. This audit, conducted in August 2007, indicated that AQSIQ's certification that the firm meets the requirements of FDA's Seafood HACCP Regulation with respect to the control of aquaculture drugs was well founded.
2. FDA reviewed reports of third party laboratory analysis of five consecutive shrimp shipments from the firm in which no drug residues were found.
3. FDA reviewed a certification from the Chinese government that the firm is in compliance with Chinese requirements for exporting seafood to the United States.

Based upon our review, we determined that the firm had demonstrated an ability to consistently produce a non-violative product. Therefore, on September 18, 2007, FDA removed Guolian from detention without examination under Import Alert 16-131. The Agency has reviewed reports of third party laboratory analysis from 27 shrimp shipments produced by the firm from the June 28, 2007 effective date of the Import Alert to the present. No unapproved drug residue was found in any of these samples.

Approximately 26 Chinese firms have requested their removal from Import Alert 16-131. To date, none have fully met FDA's expectations for removal from the Import Alert and Guolian remains the only firm exempted. However, AQSIQ has certified compliance for an additional thirteen Chinese firms and FDA has completed its paper review of the materials related to these firms. An on-site audit of AQSIQ's certification inspections of these firms is planned for July 2008. Successful completion of the audit will result in removal of these thirteen firms from the Import Alert and acceptance of future AQSIQ certification for other firms as meeting the second element (listed above) of removal from the Import Alert.

The National Marine Fisheries Service has also submitted a list of firms that they have certified to be incompliance with the Seafood HACCP Regulation with respect to the control of aquaculture drugs. FDA is awaiting materials from these firms to complete a paper review and, ultimately, an on-site audit.

MEMORANDA OF AGREEMENT

In September 2006, President Bush and Chinese President Hu Jintao agreed to create a Strategic Economic Dialogue between the United States and China. Reflecting the growing relationship between the U.S. and Chinese economies, the Strategic Economic Dialogue is designed to be a forum for discussing ways the United States and China can work together to address economic challenges and opportunities as responsible stakeholders in the international economic system. Last May, in conjunction with the 2nd Strategic Economic Dialogue, FDA and others within the Department of Health and Human Services (HHS) initiated discussions regarding the need for stronger agreements with relevant regulatory agencies in China. The agreements are intended to help assure the safety, quality and effectiveness of FDA-regulated products exported from China to the U.S.

In early December 2007, HHS Secretary Leavitt and Chinese officials signed agreements that represent an unprecedented advance in FDA's efforts to ensure the safety and quality of food, feed and medical products imported from China. The documents, which were countersigned by the Minister of the General Administration of Quality Supervision, Inspection and Quarantine (AQSIQ) and the Commissioner of the Chinese State Food and Drug Administration (SFDA), create an incremental, confidence-building system for enforcing compliance with U.S. standards in regulated products before they leave China.

The AQSIQ agreement requires specific actions to be carried out by clear deadlines, and is based on a three-pronged strategy of registration, certification and verification. First, the Chinese manufacturers of the agreed-upon items have to register with AQSIQ. AQSIQ will share the registration data with FDA, and the food producers must agree to regular inspections to ensure their exports meet the U.S. standards. Second, the agreement specifies that AQSIQ will certify food and feed covered by the agreement that meets FDA's standards. Third, to verify compliance, the Chinese are adopting quality-assurance methods to be applied at each step of the production process. For example, Chinese authorities will develop a comprehensive electronic tracking system to follow products from production to exportation. This will help ensure that growers and manufacturers are building quality into their processes and that FDA can take action if they do not.

Another critical aspect of these agreements is information sharing. Chinese authorities have pledged to provide timely notification to U.S. regulators under a wide range of circumstances, including the failure of a facility to meet inspection requirements and the suspension or revocation of a manufacturer's certification status. FDA inspectors will also gain broader access to Chinese production facilities and on an expedited basis.

FOOD PROTECTION PLAN

Going forward, all of FDA's seafood safety efforts will be informed by the Food Protection Plan announced in November 2007. The Food Protection Plan defines a science and a risk-based approach to better ensure the safety of domestic and imported foods eaten by American consumers.

The Plan, which focuses on both domestic and imported food, complements the Import Safety Action Plan that describes how the U.S. can improve the safety of all imported products. The Import Safety Action Plan lays out a road map with short- and long-term recommendations to enhance product safety at every step of the import life cycle. Taken together, the two plans will improve efforts by the public and private sector to enhance the safety of a wide array of products used by American consumers.

The Food Protection Plan is premised on preventing harm before it can occur, intervening at key points in the food production system, and responding immediately when problems are identified. Within these three overarching areas of protection, the plan contains a number of action steps as well as a set of legislative proposals. Taken together, these efforts will provide a food protection framework that ensures that the U.S. food supply remains safe.

To strengthen its efforts to *prevent* contamination, FDA plans to strengthen support of food industry efforts to build safety into products manufactured either domestically or imported. The Agency will work with industry, state, local, and foreign governments to identify vulnerabilities and will look to industry to mitigate those vulnerabilities, using effective methods such as preventive controls.

The plan's *intervention* element emphasizes focusing inspections and sampling based on risk at the manufacturer and processor level, for both domestic and imported products, that will help verify the preventive controls. This approach is complemented by targeted, risk-based inspections at the points where foreign food products enter the United States.

The plan calls for enhancing FDA's information systems related to both domestic and imported foods to better *respond* to food safety threats and communicate during an emergency.

The Food Protection Plan's three core elements--prevention, intervention, and response-- incorporate four cross-cutting principles for comprehensive food protection along the entire production chain by:

- Focusing on risks over a product's life cycle from production to consumption;
- Targeting resources to achieve greatest risk reduction;
- Using interventions that address both food safety (unintentional contamination) and food defense (deliberate contamination); and
- Using science and employing modern technology, including enhanced information technology systems.

CONCLUSION

Ensuring the safety of the food supply continues to be a top priority for FDA and we are working hard to ensure the safety of all human food and animal feed, in collaboration with our Federal, state, local, and international food safety partners. FDA is working diligently to efficiently and effectively use the resources and authorities provided by Congress to protect the public health of the U.S. and to help ensure that imported products are safe for American consumers. Despite the challenges which face us, the American food supply continues to be among the safest in the world. Thank you for the opportunity to provide testimony to the Commission.

In: Seafood Fraud and Safety: Background and... ISBN: 978-1-61761-500-9
Editors: Kaitlyn O. Harris

Chapter 7

TESTIMONY OF E. SPENCER GARRETT, NATIONAL OCEANOGRAPHIC AND ATMOSPHERIC ADMINISTRATION, U.S. DEPT. OF COMMERCE, BEFORE THE U.S.-CHINA ECONOMIC AND SECURITY REVIEW COMMISSION, HEARING ON "NOAA'S ROLE IN SEAFOOD SAFETY MONITORING OF CHINESE-ORIGIN PRODUCTS"

Good morning, I am E. Spencer Garrett, Director of the National Seafood Inspection Laboratory located in Pascagoula, Mississippi. Our Laboratory is part of the National Oceanic and Administration (NOAA) National Marine Fisheries Service (NMFS), which is under the U.S. Department of Commerce (DOC).

NOAA's National Marine Fisheries Service is responsible for the stewardship of the nation's living marine resources within the world's largest exclusive economic zone (EEZ). NOAA/NMFS protects and conserves these resources and their habitats through scientific research, fisheries management, efforts to protect marine mammals and endangered species, law enforcement, habitat conservation, and voluntary seafood quality and safety monitoring and inspection. NMFS has both domestic and international responsibilities, and seeks to maximize economic benefits from the sustainable use and conservation of living marine resources.

The principal points I will be discussing this morning relate to the complexities of the U.S. fishery system. I will also provide a summary of how the NMFS National Seafood Inspection Program (NSIP) assists and complements the activities of the U.S. Food and Drug Administration (FDA) in addressing consumer hazards in the consumption of seafood products, including those from China.

U.S. FISHERY SYSTEM

First, I would like to describe the U.S. fishery system from a domestic viewpoint. I'll begin with our environmental fishery habitats, which must be protected so the United States will have a recurring, sustainable fishery resource. Our coastal estuaries serve as a breeding ground providing habitat for more than 75 percent of commercial landings and 80 to 90 percent of recreational catch of fish and shellfish. From these habitats, hundreds of species of seafood are produced. In addition to wild species, aquacultured species (including imports) now contribute up to 45 percent of the U.S. seafood supply. Our wild species are harvested by 12 million recreational anglers and nearly 300,000 commercial harvesters. To give you an idea of the magnitude of marine recreational fishing in the United States, in 2006 recreational fishermen made approximately 89 million recreational fishing trips catching 476 million fish, of which more than half were released. In 2000, fishermen spent $13.6 billion, which translates into more than $31 billion in sales for the U.S. economy. The commercial harvesters deploy 90,000 vessels, while recreational fishermen operate hundreds of thousands of recreational fishing boats. There are nearly 5,000 domestic plants and wholesalers located in every state, not just among our coastal areas as commonly perceived. Our per capita consumption of commercially harvested species averaged 16.5 pounds in 2006. It is estimated that our per capita consumption of recreationally harvested seafood approaches an additional 3 to 4 pounds. Our country, as well as other highly industrialized nations, does not solely rely on domestically produced products, but rather must import a large portion of seafood products for consumption. For a number of years, over half of our seafood consumption has relied on foreign produced products. Currently, we import over 80 percent of our total consumption of seafood products. This trend for economic reliance on seafood imports has been steadily increasing over the past 10 years to the extent that the United States is now the world's second largest importer of seafood. When dealing with imports, it is often difficult to determine exactly where the seafood was harvested; the origins of imported seafood may have been masked.

U.S. participation in the international trading of seafood is complex; not only are we the world's second largest importer of seafood, but we are also the world's fourth largest exporter of such commodities. This dichotomy of circumstances complicates our participation in international trading of seafood when developing marketing and import/export food control inspection strategies.

As the amount of seafood imported into our country has grown, the safety of these products has become a critical concern. The United States has an annual $9.1 billion trade deficit with respect to seafood. In 2006, we imported 1.2 billion pounds of seafood from China valued at approximately $1.9 billion. In contrast, we exported 500 million pounds of seafood to China valued at approximately $450 million.

SEAFOOD PUBLIC HEALTH RISK PROFILE

Traditionally, seafood product safety risks have been subcategorized into four categories: environmental, process, distribution, and consumer induced. The environmental group is further divided into natural hazards (i.e., biotoxins and marine pathogenic bacteria), or anthropogenic contaminants (i.e., PCBs, pathogens, antibiotics, residues, etc.).

HOW SAFE IS SEAFOOD?

Consumer hazards in the consumption of all food, including seafood, can be grouped into three areas: product safety, food hygiene (i.e., clean vs. dirty processing plants, wholesome vs. unwholesome products, etc.), and mislabeling or economic fraud (which can include allergenic considerations).

Let's focus for a moment on the often asked question, "How safe is seafood?" The question has been answered many times by the National Academy of Sciences (NAS), the Centers for Disease Control and Prevention (CDC), and others. The answer is simply that the vast majority of seafood is safe. However, as with all foods, there are some risks, and in the case of seafood the food safety issues are highly focused, well defined, and concentrated in a very few species. It has been determined, for example, that for seafood-borne illnesses reported to the CDC where the cause was actually known, more than 72 percent of the outbreaks and 38 percent of the individual cases relate to ciguatoxin (from a few reef fish species), scombrotoxin (from tunas, mackerels, bluefish, and a few other species), or the consumption of molluscan shellfish (mostly in the raw state).

How Prevalent Is Economic Fraud?

Economic fraud issues have been significant for a number of years throughout the domestic and imported seafood industry. Fraud is sometimes referred to as a "shrouded consumer hazard," as few databases are specifically designed to collect economic fraud data. When economic fraud data are reported, they are usually incidental discoveries. Economic fraud tests, when performed, are usually secondary to the investigative study. Not all occurrences of economic fraud are intentional; they can be caused by a lack of knowledge of the multiplicity of species in the U.S. market, or a lack of communication on the part of the processors, packers, or buyers. Economically fraudulent or deceptive practices within the seafood industry include: mislabeling or substitution of lower value species for higher value species, low weights or undercounting, over treating or added water weight, altered color, and transshipment of products to avoid import or customs duties or to mask the identity of either the country of origin or specific plants within a given country.

An excellent description of these fraudulent issues and federal and industry steps to curtail such activities can be found in the 2007 Congressional Research Service report, *Seafood Marketing: Combating Fraud and Deception* by Eugene Buck. One point of applicable law omitted in that report is that NMFS, through the Lacey Act and the Agricultural Marketing Act of 1946, can take direct action against mislabeled products and that our Office for Law Enforcement has taken such action. Our Seafood Inspection Program notifies the FDA when such issues are encountered or works directly with the NMFS Office for Law Enforcement.

NOAA's Seafood Inspection Program

NOAA's Seafood Inspection Program provides inspection services for the domestic and international seafood industries that directly affect American consumers. The Seafood Inspection Program is a voluntary, "fee-for-service" program that derives its legislative authority from the Agricultural Marketing Act of 1946 (7 U.S.C. 162 1-1627). The Program has been in existence for over 50 years, first within the Department of the Interior and later in NOAA. Its primary mission is to assist the seafood industry in producing high-quality and safe products for the benefit of the American consumer.

To meet this mission, the Program conducts in-plant process and sanitation evaluations, product grading and evaluations, and certification services. We consult on and implement systems designed to prevent food safety and quality problems from occurring. The Program assists with developing product standards and specifications that are consistent with applicable federal laws and regulations. All of these services are designed to help seafood processing firms produce higher-quality, safe products that comply with all applicable regulatory requirements, including FDA laws and regulations.

The primary clients of NOAA's Seafood Inspection Program are seafood processing firms, importers, and exporters. In turn, their customers are large seafood buyers such as supermarket chains and the U.S. military. In 2006, the Program inspected approximately 1.9 billion pounds of seafood, or approximately 30 percent of the seafood in the United States. These domestic inspection services were accomplished through contracts with approximately 300 participants, and many more U.S. clients use the service on an on-demand basis. Approximately 35 foreign participants, including 23 from China, also participate in the NOAA Seafood Inspection Program.

The voluntary fee-for-service Program derives all of its operating revenues from program participants. It does not use taxpayer money. The FY 2008 estimated budget for the Program is $20 million. NOAA estimates that the average cost for added quality and safety assurance provided by the Program is about $0.01 per pound.

The Program employs about 165 people, sited across the United States. It receives support from a small staff of scientific and technical experts who provide training and analytical services, develop product standards and specifications consistent with applicable federal laws and regulations, and ensure that consumer labels meet all regulatory standards.

SCIENTIFIC SUPPORT FOR NOAA'S SEAFOOD INSPECTION PROGRAM

Broad scientific support for the Seafood Inspection Program is provided primarily by the National Seafood Inspection Laboratory in Pascagoula, Mississippi, and NMFS' Seafood Product Quality and Safety Research and Monitoring Program at the NMFS Northwest Fisheries Science Center in Seattle, Washington. These two distinct programs allow NMFS to proactively and rapidly respond to seafood safety and aquatic animal health issues. We also address episodic events on multiple levels from specific fish or shellfish and human health concerns within the broader marine environment. The National Seafood Inspection Laboratory provides thousands of analyses of NOAA-inspected products to determine compliance requirements with scombrotoxin, sodium bisulfite, antibiotic residues, selected microbial pathogens, and indicator organisms.

While the Seafood Inspection Program is funded through user fees, the science and research activities at the National Seafood Inspection Laboratory and Northwest Fisheries Science Center are funded through appropriated funds.

These research and monitoring programs provide, in large part, the necessary capability for NOAA to respond quickly to environmental disasters that can affect seafood safety and quality. The value of this capability was evident in NOAA's rapid response to possible human health threats in the aftermath of Hurricane Katrina. Without this support, NOAA's rapid response to Hurricane Katrina would not have been possible. The availability of trained and experienced staff and laboratory analytical capability is a critical prerequisite for a rapid, timely, and effective response to all seafood safety issues. These research and monitoring program activities, following Hurricane Katrina, helped FDA to ensure public confidence in the safety of the $7 billion Gulf of Mexico seafood industry.

The NMFS Northwest Fisheries Science Center uses state-of-the-art research facilities to assess a wide range of chemical contaminants, pathogens, and marine toxins in seafood samples. The Center also has biological knowledge and experience with emergency field response, having responded to seafood safety issues after the Exxon Valdez oil spill in 1989. Most recently, the Center responded to the public's concern over seafood safety following Hurricane Katrina, when many laboratories in the storm-affected area were damaged and/or inoperable. The Center worked collaboratively with

FDA and other agencies. NOAA sent a team to the affected area within days of the storm, and within weeks it provided precise technical information which helped to verify the safety of seafood from the northern Gulf of Mexico.

Relationship between the Seafood Inspection Program and the Food and Drug Administration

NMFS has a long history of working cooperatively with the FDA regarding seafood quality and safety risk assessment, management, and communication issues. We cooperate from both an inspection and certification perspective, as well as from a fishery science, seafood safety research, and monitoring viewpoint.

With regard to Chinese seafood imports, an important part of the mission of the FDA is to protect the safety of food for the American consumer. The NOAA Seafood Inspection Program uses FDA criteria to evaluate products and processes.

About 83 percent of all seafood consumed in the United States comes from imports, a volume of approximately 4.1 billion pounds annually in 2006. China, Canada, Thailand, and Chile are the principal countries of origin for imports, with China shipping about 1.2 billion pounds to the United States per year. This volume demonstrates the challenges of effectively regulating and evaluating the seafood flowing into the United States.

The Seafood Inspection Program has been meeting with FDA to update the current Memorandum of Understanding between the agencies on seafood inspection matters in order to better serve the American consumer.

Conclusion

Mr. Chairman, Commissioners, and staff, NOAA and NMFS look forward to working with you, the public, the seafood industry, and the FDA to ensure the safety of the seafood we consume. I will be happy to answer any of your questions.

In: Seafood Fraud and Safety: Background and... ISBN: 978-1-61761-500-9
Editors: Kaitlyn O. Harris

Chapter 8

TESTIMONY OF PATRICK WOODALL, SENIOR POLICY ANALYST, FOOD & WATER WATCH, BEFORE THE U.S.-CHINA ECONOMIC AND SECURITY REVIEW COMMISSION, HEARING ON "CHINESE SEAFOOD IMPORTS: SAFETY AND TRADE ISSUES"

Good morning Co-Chairs Bartholomew and Slane and members of the Commission. My name is Patrick Woodall and I am a Senior Policy Analyst at Food & Water Watch, a non-profit consumer organization that works on food policy and water infrastructure issues. We were founded in November 2005 and we are based in Washington, D.C.

Thank you for holding this hearing and inviting Food & Water Watch to testify on the food safety implications of imported Chinese seafood. China is a fisheries superpower and leads the world in total fish catch, fishing fleet size, aquaculture production and global seafood exports. China's fishing and aquaculture industry can have significant environmental impacts on the global health of ocean fisheries, China's environmental quality and the safety of seafood for consumers in China and here in the United States. China's vast aquaculture production poses special risks to food safety, since densely packed fish are often treated with antibiotics and veterinary medicines that are illegal to use on fish in America because of their associated human health and environmental risks.

China's seafood exports to the United States have nearly tripled in recent years, from 363 million pounds in 2000 to nearly a billion pounds in 2007. Shipping seafood 6,500 miles from China to California can increase the likelihood of microbial contamination and decomposition during storage and transit. This steep increase in the volume of imported fish has rapidly exceeded our food safety oversight at the border. The FDA physically inspects fewer than two percent of imported fish shipments and only refused entry to one out of 476 shipments in 2006. The capacity of our fish import inspection system has not grown to meet the increased demands rising imports have placed upon America's food safety infrastructure or personnel. The number of inspectors, inspections and laboratory tests needs to rise to the challenge of escalating seafood imports.

More fish facing fewer inspections has real implications for American consumers. As much as 84 percent of fish and shellfish consumed in the United States is imported and 40 percent of imports come from fish farms.[1] Seafood products cause about a fifth (between 18 and 20 percent) of the known outbreaks of foodborne disease in the United States each year. All cases of foodborne illness lead to about 325,000 hospitalizations and 5,000 deaths annually in the United States.[2]

The safety of products from China received considerable attention in 2007. Import scandals over tainted pet food ingredients, industrial solvent-laced toothpaste and lead painted toys focused needed attention on whether U.S. regulators can effectively monitor a tidal wave of imports from China and the rest of the world. Over the past several years, Japanese, European and American state regulators have acted to prohibit dangerous Chinese seafood imports from reaching grocery stores. The FDA finally imposed a country-wide ban on some Chinese seafood products in the summer of 2007, when it banned five types of aquaculture produced fish that frequently tested positive for veterinary medicines that are not approved for use on fish in the United States.

The sheer volume of fish imports from China overwhelms our current capacity to ensure the safety of the fish Americans eat. The magnitude of China's aquaculture industry and its reliance on banned chemicals as well as China's weak enforcement of food and environmental regulations could continue to pose a risk to American consumers. In October 2007, the Subcommittee on Oversight and Investigations of the U.S. House Energy and Commerce Committee investigated the safety of China's food exports and reported, "It would appear that the Chinese food supply chain does not meet international safety standards. It is, in fact, responsible for very serious domestic Chinese food poisoning outbreaks."[3]

China is World's Largest Fish Producer

China is the global fishing leader and is a production powerhouse from its fish farms to its industrial fishing fleet. In 2004, China produced 105 billion pounds of fish and seafood – 37 billion pounds from capture fisheries and 68 billion pounds from aquaculture production.[4] China's capture fishery production nearly doubled from fewer than 20 billion pounds in the mid-1980s to 38 billion pounds in 1998, or about one-fifth of global fish production.[5] In 2004, China's 37.3 billion pounds of capture fishery production made China the world's largest capture fishing industry – 75 percent higher than the second place Peru and three times larger than the United States, which ranks third globally.[6] China flagged more than half a million powered fishing ships in 2005, making it the largest fleet as well.[7]

The tens of billions of pounds of fish landed by China's fishing fleet and its disproportionate impact on the global wild-capture fisheries threaten the very survival of ocean fish species. China's vast fishing fleet contributes both to local and global overfishing.[8] A 2006 study in *Science* found that current fishing trends could lead to a global collapse of all species of marine food fish by the middle of this century.[9] The UN reports that three quarters of the most important food fish species are already fully or overexploited.[10]

China's Unrivaled Aquaculture Industry

Aquaculture is believed to be the fastest growing food-producing sector in the world and constitutes nearly half of all consumed fish.[11] China is the global leader in aquaculture production as well. China's aquaculture production increased by 12.4 percent annually between 1950 and 2004.[12] China's aquaculture industry is the largest in the world and produced 91 billion pounds of fish in 2004, 69.6 percent of global aquaculture production.[13] More than three-fifths (64 percent) of China's fish production comes from aquaculture, making China the only nation where aquaculture production exceeds wild-caught fish.[14] China was the top aquaculture producer of tilapia (1.9 billion pounds) and sixth largest producer aquaculture producer of mullet (1.2 million pounds) in 2004.[15]

China's aquaculture sector includes large industrial facilities as well as millions of small- scale fish farms, making it difficult to ensure the safety and quality of the fish. The state Fishery Bureau estimates there are 4.5 million fish farmers in China – a 20 percent increase in aquaculture employment since 2000.[16] The majority of China's food and fish production is on small-scale family farms.[17] These small, rural, family-run fish farms can apply fish feed and chemicals to treat the water, but the farmers may not know what they are adding to their ponds.[18] They buy these often unlabeled inputs from salespeople that promise increased fish production and growth. Thousands of fish processing plants buy fish from these small farmers for the Chinese consumer and U.S. export market.[19] Retired Food and Drug Administration associate commissioner William Hubbard observed that food production in China "is a vast cottage industry," and the central Beijing authorities have "very little influence" on the fragmented food industry.[20] The Wild West capitalism that exists in China makes it difficult to enforce even the most well intentioned food safety rules. The melamine tainted pet food incident and more recent heparin contamination revealed a complex network of small producers that evaporated when Chinese regulators attempted to get to the bottom of the contaminations.

Aquaculture production can have unique food safety implications and environmental impacts. Fish grown on farms require clean water to be healthy. Densely packed fish raised in polluted water can be contaminated with industrial toxins and susceptible to disease. China's polluted waterways threaten the safety of the country's aquaculture products. Industrial pollution and municipal sewage create serious water pollution problems for China's aquaculture.[21] According to the Chinese Fisheries Bureau's Director of Aquaculture "Water quality is the top issue for Chinese aquaculture."[22] Some aquaculture facilities raise fish in water that is polluted with sewage, farm runoff containing pesticides and industrial chemical discharges.[23] In southern China, consumers that have eaten DDT-laced fish are facing serious health problems.[24] Mercury pollution from China's coal-fired power plants also contaminates China's waterways and potentially its aquaculture industry.[25] China's weak and under-enforced environmental controls allows these water pollutants to contaminate aquaculture facilities and increases the risk of microbiological, metal and toxic contamination of aquaculture produced fish products.[26]

Aquaculture producers in China have tried to combat pollution-related bacterial infections, disease and parasites by applying antibiotics, fungicides and other pesticides. Some of these antibiotics (notably malachite green, nitrofurans, flouroquinolones and chloramphenical) may contribute to increased antibiotic resistant strains of bacteria that infect humans. Several other chemicals (nitrofurans, malachite green and gentian violet) have been associated with increased human cancer risks.[27] Some veterinary drugs are prohibited in livestock and aquaculture production because their use can facilitate antibiotic resistant foodborne pathogens. Since there are not systems in place to monitor antibiotic resistant foodborne pathogens in fish products, the FDA has prohibited their use in aquaculture production.[28] The application of these veterinary medicines affects not only the fish they aim to treat, but the surrounding environment and non-target species, which can cause anti-microbial resistance and impact human health.[29] Although China banned malachite green in 2002, it continues to allow the use of fluorquinolones in aquaculture.[30] The FDA has continued to find these chemicals in imported aquacultureproduced fish despite China's official ban.

Aquaculture production also contributes to water pollution, global fish depletion and can threaten native species. In some areas in China, aquaculture ponds are packed so densely that their effluents pollute downstream fish farms.[31] The runoff and wastewater from aquaculture production contains uneaten food, fish waste and agrochemicals that can produce considerable effluents that can feed algae blooms and eutrophication of waterways, depriving native species of necessary oxygen in the water and creating dead zones in coastal areas.[32] China has already started to curtail or ban aquaculture production at some locations because of concerns about water pollution from fish farms. In 2007, a provincial government banned crab farming at the Yangcheng Lake to protect the drinking water supply from algae blooms created by excessive high-nutrient effluent from the aquaculture operations.[33] All aquaculture production was removed from China's largest three lakes this year to prevent future economic damage from algae blooms.[34]

Additionally, farm raised fish primarily eat fish feed that mostly ground fishmeal from small ocean fish that are a critical part of marine ecosystems. It can take between two to six pounds of small, ocean fish to produce one pound of farmed fish.[35] More than a third (37 percent) of all seafood is ground into fishmeal, and a third of that fishmeal is consumed by Chinese aquaculture.[36]

Finally, the introduction of non-native species for aquaculture can also threaten existing ecosystems, as the Nile perch came to dominate Lake Victoria, and can introduce disease to native species, as European brown trout introduced Whirling disease to the U.S. rainbow trout population.[37] Escapes are inherent to the aquaculture industry, and since most farmed fish lack significant genetic diversity, escaping fish can pass on any genetic weaknesses to the wild fish population.

These are not solely Chinese dilemmas; most of these problems plague aquaculture and industrial fishing in other parts of the world. However, since China is a market leader, its fisheries and aquaculture model is exported to other countries that adopt China's fishing practices.

CHINA'S FISH PRODUCTION FUELS EXPORTS

China's aquaculture facilities and fishing fleet feed Chinese consumers and provide an engine for exports. China's fish and seafood exports more than tripled over the past decade and China has been the world's largest fish exporter since 2002.[38] In 1994, China exported about $2 billion in seafood products; by 2004, China exported more than $6 billion in seafood products.[39] Between 1994 and 2004, China's share of the value of global fish exports nearly doubled from 4.9 percent of all exported fish to 9.3 percent in 2004.[40] China's fish exports represented nearly a third (29 percent) of its agricultural exports in 2004.[41]

The United States is the third largest importer of fish products, bringing in 16 percent of the value of all global fish trade.[42] China is the third largest supplier of these fish imports and China's fish exports to America have been growing. The value and volume of China's fish and seafood exports to the United States has grown substantially over the past two decades but has sharply accelerated since 2000, when China received permanent trade access to the U.S. market.[43]

The value of fish and seafood products from China grew rapidly and much more quickly than fisheries exports from the rest of the world. Between 1989 and 1999, the value of U.S. fish imports grew an average of about 5 percent a year, and the import growth for Chinese imports mirrored the growth for the rest of the world (4.7 percent and 5.0 percent, respectively). During this period, China's seafood exports to the United States grew from $297 million in 1989 to $357 million in 1999 (see appendix Chart 1). After 2000, the value of China's fish exports to the United States grew by 20.8 percent a year – more than five times greater than the 3.6 percent annual growth for all other fish exporters. In the last seven years, Chinese fish exports nearly tripled from

$521 million in 2000 to $1.5 trillion in 2007. This large increase in fish export sales also increased China's market share of the U.S. imported seafood market, growing from below 5 percent of sales in the mid-1990s to 14.5 percent of all fish imports in 2007 (see Chart 2).

The dollar value of exports demonstrates the muscle of China's fishing industry but does little to describe the magnitude of the problem for U.S. import inspectors. It is easier to examine $1 million worth of lobster than $1 million worth of anchovies. The tonnage of Chinese imports has grown steadily over the past two decades, rising rapidly since 2000. U.S. inspectors faced nearly a billion pounds of Chinese seafood imports in 2007. Between 1989 and 1999, the volume of China's fish exports to the United States grew at an annual average rate of 10.9 percent, from 118 million pounds in 1989 to 288 million pounds in 1999 (see Chart 3). Since 2000, the volume of fisheries exports grew by 16.8 percent annually from 363 million pounds in 2000 to 945 million pounds in 2007. China's share of U.S. fish and seafood imports has quintupled in the past two decades. In the early 1 990s, China's fish exports represented about 5 percent of the total tonnage of fish the United States imported; by 2007, nearly one-quarter (24.7 percent) of imported fish came from China (see Chart 4).

U.S. IMPORT INSPECTION SYSTEM UNLIKELY TO SPOT PROBLEMS IN FISH IMPORTS

The FDA is responsible for ensuring the safety of U.S. fish imports, but the increasing volume of imports is overseen by an inspection regime that has not grown to meet the challenge and leaves American consumers poorly protected. Less than one out of fifty shipments of fish received a physical inspection at the U.S. border between 2003 and 2006.[44] Physical inspection can include document reviews, sensory evaluations, label examinations or laboratory testing. In 2006, about one in a hundred (1.3 percent) of imported fish shipments received a sensory examination and one in two hundred (0.59 percent) received a laboratory inspection. Laboratory testing of seafood has declined from 0.88 percent of shipments in 2003 to 0.59 percent in 2006 – a 33 percent decline.

The level of import inspections by the FDA is significantly below other industrialized countries that import large volumes of seafood products. The European Union inspection rate is ten times higher and Japan's inspection rate is more than five times higher than the U.S. inspection rate for seafood. The European Union inspects 20 percent of fish that is fresh, frozen, dried, salted or packed in airtight containers and 50 percent of bivalves like oysters and scallops.[45] Japan tested 12 percent of all seafood products and 21 percent of processed seafood in 2005.[46]

FDA'S REFUSALS OF FISH IMPORTS FROM CHINA RISING

The overall number and proportion of U.S. seafood imports that are rejected by border inspectors is very small. The U.S. rejected 1,786 shipments of fish in 2006, more than a third less than the 2,693 that were rejected in 2003. The percentage of shipments of seafood that were refused has been declining for the past few years. In 2003, the FDA refused 0.36 percent of seafood shipments but in 2006 the FDA refused 0.21 percent – a 42 percent decline.

While the number and percentage of refused seafood shipments has been declining, Chinese seafood refusals are a growing share of all rejected seafood and the number of refusals from China has been growing. The share of refused seafood shipments from China has nearly tripled from 6.8 percent of all seafood refusals in 2003 to 17.3 percent of all refusals in 2006. Between 1997 and 2006, the FDA refused 1,364 shipments of fish and seafood products from China for failing to meet U.S. standards for import (see Chart 5).[47] In more recent years, FDA refusals of Chinese fish imports have increased. Between 1997 and 2001, the FDA refused an average of 76 shipments each year, but between 2002 and 2006 the FDA refused an average 196 shipments per year. Between 2005 and 2006 alone, the number of FDA refusals increased by more than 50 percent from 198 in 2005 to 309 in 2006.

These border refusals were predominantly related to significant safety concerns. Nearly a sixth of the refusals were for foodborne pathogens on Chinese fish – 13 percent of refusals were for *Salmonella* and three percent were for *Listeria*. Nearly one-fifth (17 percent) of the refusals were to stop fish with illegal veterinary drug residues. Nearly a third (32.5 percent) of the refusals were for fish that the FDA deemed too filthy for human consumption. Nearly one in twenty refusals were for unsanitary fish shipments (2.5 percent of refusals) or for findings of pesticides, histamines, dangerous color additives, or contamination with heavy metals like mercury (2.1 percent combined).[48] (See Chart 6.)

Some common aquaculture produced fish from China had high levels of refusals for illegal veterinary drugs and antibiotics. Almost all (88.1 percent) of Chinese eel refusals were for veterinary drug residues between 2005 and 2006. Three out of five (63.8 percent) of Chinese catfish refusals in 2006 were rejected for veterinary drugs. More than half (51.9 percent) of Chinese refused crab imports between 2003 and 2006 were for illegal veterinary drug residues. More than two-fifths of refused shrimp (42.4 percent) from China between 2002 and 2006 and tilapia (45.5 percent) rejected between 2005 and 2006 were rejected for veterinary drug residues. (Testing for veterinary medicine did not become common until after 2001, and for many types of fish products veterinary drugs were not tested for until 2005.)

Critical shortages of inspectors at FDA, where the overall food inspection rate is below one percent of shipments, indicate more resources for inspection are needed. Replacing or supplementing inspectors with third party certification is no solution to the weak oversight by the FDA. Food inspection is a basic government function. Relying on third party certifiers would essentially privatize food inspection. It is simply not adequate for the government to "verify" third party inspectors through paperwork checks. That is not government inspection; that is government auditing of paperwork.

A July 2007 Energy and Commerce Committee Subcommittee on Oversight and Investigations staff report found that the third party laboratories that currently certify import safety force the FDA to rely on unverified laboratory findings to allow imports to proceed to U.S. grocery stores and the quality of third party testing operations varies widely.[49] One FDA laboratory deputy director deemed some of the third party lab work as "scary,' "not good," and "spooky." Another viewed the third party laboratory work as "shoddy" because the private companies had a financial interest in clearing the imports.

Growing International Concern over China's Aquaculture Safety Culminates in 2007 FDA Import Ban on Certain Aquaculture Produced Fish

China's fish exports, especially aquaculture-produced fish, have faced increased regulatory oversight as more authorities have found banned chemical residues on imported fish. The FDA has had some import restrictions on certain fish from certain Chinese companies since 2001 because of the presence of veterinary drugs, antibiotics and other additives that are prohibited for use on fish in the United States.[50] In 2003 the EU banned Chinese shrimp and in 2005 Japan stopped Chinese eel imports.[51] Taiwan banned the import of certain crab types in 2006 after finding cancer-causing veterinary drug residues that were 16 times higher than Taiwan's permitted residue levels; Taiwan estimated that consumers ate more than 6,600 pounds of contaminated crabs before the ban went into effect.[52] Even China pulled aquaculture-raised fish off the market in recent years. In November 2006, China banned the sale of turbot after finding that 100 percent of turbot sampled in Shanghai tested positive for carcinogenic veterinary drugs.[53]

Although the United States banned the imports of fish from some Chinese fish processors with a history of problems, companies could apply to have their ban lifted by appealing the ruling and successfully submitting five shipments to laboratory testing. Despite increased international attention to the issue of contaminants in Chinese fish, the FDA did not extend its oversight to all Chinese fish until 2006. FDA Assistant Commissioner David Acheson noted at a press conference after the FDA imposed a ban on some Chinese imports in 2007 that there was "evidence of problems [with illegal veterinary drug residues] over a period of years" including "extensive communications" between U.S. and Chinese regulators, but veterinary drug residues persisted in Chinese imports.[54] The FDA finally started increased sampling of imports in 2006.[55]

In the United States, increased scrutiny of Chinese fish products followed the 2007 fallout over the surge of dangerous products from China ranging from tainted pet food ingredients to lead-tainted toys. In April 2007, Wal-Mart removed all frozen catfish fillets from China when Alabama banned Chinese catfish sales after finding banned antibiotics on Chinese imports.[56] In May 2007, the Mississippi Commissioner of Agriculture ordered some stores to stop selling Chinese catfish imports after finding residues of antibiotics banned in the United States.[57] Congressional investigations and consumer group attention to this issue, including the Food & Water Watch *Import Alert* study in 2007, focused national attention on the risks posed by fish imports, especially from China's aquaculture industry.

In June 2007, the FDA banned the import of all Chinese shrimp, catfish, dace, basa and eel because of long-term problems with illegal veterinary drug residues. The FDA's import alert required firms to affirmatively demonstrate the safety of their product before their imports would no longer be detained without examination, but the first firm that was exempted from the alert has already been red-flagged by Canadian authorities for veterinary drug residues. The first Chinese seafood company was exempted from the FDA Import Alert in September 2007, three months after the country-wide ban went into effect, after the FDA inspected the company's processing plant and five consecutive export shipments were cleared by third-party analysis.[58] Once the Import Alert exemption was in place, the exporter's fish were inspected only as frequently as other fish imports, but a month after the FDA lifted the alert on the Chinese company, Canada issued its own Import Alert after finding a cancer causing antibiotic in a shipment of frozen shrimp from the company.[59]

Even with the FDA Import Alert in place, Chinese exports could evade the system by transshipping their fish through third countries. Under U.S. law, exporting countries can legally re-route their goods and products through intermediary countries, if that transshipment is not used to circumvent either customs duties, such as import protection, anti-dumping or countervailing duty orders, or trade restrictions such as FDA Import Alerts on dangerous fisheries products. [60] Illegal fishing in excess of permits, treaties or quotas often relies on industrial trawlers with false flags and national registrations which offload their catches at sea and re-route their catch through ports with few resources or inspectors, which effectively launders their illegal catch.[61] Food & Water Watch is concerned that some fishery and seafood exporters could similarly circumvent import safety restrictions (like the ones currently in place on some Chinese aquaculture products) by transshipping fish through third party ports.

There are indications that Chinese fish exporters may already be transshipping fish products to evade current antidumping orders. In 2006, the Indonesian Minister of Marine Resources and Fisheries banned seven Indonesian fishery products exporters from shipping shrimp to the United States after finding that Chinese shrimp labeled as Indonesian products were being shipped to the United States.[62] China's exports to Hong Kong – of fish, seafood and other products – may not be going to the Hong Kong consumer market. Many of these goods are re-routed in Hong Kong port facilities to a second, re-exported destination. A white paper from the U.S. International Trade Commission last month reports that Chinese exporters and Hong Kong re-exporters do not know the final destination of goods exported from China until they clear Hong Kong's customs as imports.[63]

CONCLUSION

There are not enough resources or inspectors to sufficiently monitor Chinese seafood imports. The FDA needs to increase physical inspection of imported seafood at least to the levels used in the European Union or Japan. Laboratory inspections should be focused on aquaculture products that are more likely to be contaminated with antibiotics and veterinary drugs that are not approved in the United States. FDA also needs to conduct more inspections at foreign production and processing facilities. Finally, the FDA should more proactively address emerging health risks when hazardous trends are identified. The United States is the third largest seafood importer in the world and should be a flagship of import safety.

End Notes

[1] Becker, Geoffrey S., Congressional Research Service, "Food and Agricultural Imports from China," RL34080, July 17, 2007 at 2; NOAA estimate, available at www.nmfs.gov/mediacenter/aquaculture/.

[2] U.S. Centers for Disease Control, "Surveillance for Foodborne-Disease Outbreaks—United States, 1993- 1997," *Morbidity and Mortality Weekly*, 49(SS-1), March 1997.

[3] U.S. House Energy and Commerce Committee, Subcommittee on Oversight and Investigations, Staff Trip Report, "Food From China: Can We Import Safely," October 4, 2007 at 1.

[4] United Nations Food and Agriculture Organization, *The State of World Fisheries and Aquaculture 2006*, 2007 at 4.

[5] Allain, Marc, Greenpeace International, "Trading Away Our Oceans," January 2007 at 14.

[6] *The State of World Fisheries and Aquaculture 2006* at 9.

[7] *Ibid.* at 26-27.

[8] Guilin, Dai and Zhou Jing, Ocean University, Qingdao, "The Basic Trend of Chinese Seafood Resources Development Strategy," Institute for Research in Economics and Business Administration (Norway), SNF Report No. 20/05, 2005 at 29-3 0.

[9] Worm, Boris, Edward B. Barnier, et al., "Impacts of Biodiversity Loss on Ocean Ecosystem Services," *Science*, November 3, 2006.

[10] *The State of World Fisheries and Aquaculture 2006*, at 29.

[11] UN Food and Agriculture Organization, *State of World Aquaculture 2006*, FAO Fisheries Technical Paper 500, 2006 at 1.

[12] *Ibid.* at 7, Table 1.

[13] *Ibid.* at 5.

[14] *The State of World Fisheries and Aquaculture 2006*, at 1.

[15] *State of World Aquaculture 2006*, at 7, Box 1.

[16] *The State of World Fisheries and Aquaculture 2006*, at 24.

[17] Ryan, Missy, "China Food Safety Seen Mired in Economic Transition," *Reuters*, December 18, 2007.

[18] Schmit, Julie, Calum MacLeod, Elizabeth Weise and Barbara Hansen, "Chinese Fish Crisis Shows Seafood Safety Challenges," *USA Today*, June 28, 2007.

[19] *Ibid.*

[20] Cha, Ariana Eunjung and Annys Shin, "U.S. Gains More Access to Data on Chinese Exports," *Washington Post*, December 12, 2007.
[21] Schmit, MacLeod, Weise and Hansen, *USA Today*, June 28, 2007.
[22] Barboza, David, "China Says Its Seafood Is Safer," *New York Times*, January 17, 2008.
[23] *Ibid.*
[24] Barboza, David, "In China, Farming Fish in Toxic Waters," *New York Times*, December 15, 2007.
[25] Ellis, Linden J. and Jennifer L. Turner, Western Kentucky University, China Environment Health Project Research Brief, "Aquaculture and Environmental Health in China," May 7, 2007 at 1.
[26] Congressional Research Service at 11.
[27] U.S. Food and Drug Administration, Import Alert 16-131, "Detention without Examination of Aquacultured Catfish, Basa, Shrimp, Dace, and Eel Products from the People's Republic of China Due to the Presence of New Animal Drugs and/or Unsafe Food Additives," August 3, 2007 Revision.
[28] FDA, Transcripts of FDA Press Conference on Seafood Imported from China," June 28, 2007 at 19.
[29] World Bank, Agriculture and Rural Development, "Aquaculture: Changing the Face of the Waters," Report No. 36622-GLB, 2006 at 20.
[30] FDA, Import Alert 16-131.
[31] Barboza, *New York Times*, December 15, 2007.
[32] Linden and Turner, "Aquaculture and Environmental Health in China," May 7, 2007 at 2.
[33] Ellis, Linden J. and Jennifer L. Turner, Woodrow Wilson International Center for Scholars, "Surf and Turf: Environmental and Food Safety Concerns of China's Aquaculture and Animal Husbandry," *China Environment Series*, Issue 9, 2007 at 19.
[34] *Ibid.* at 22.
[35] Report of the Marine Aquaculture Task Force, "Sustainable Marine Aquaculture: Fulfilling the Promise; Managing the Risks," Takoma Park, MD, January 2007 at 93.
[36] Stier, Ken, "Fish Farming's Growing Dangers," *Time*, September 19, 2007.
[37] World Bank at 21-22.
[38] *The State of World Fisheries and Aquaculture 2006*, at 7.
[39] *State of World Aquaculture 2006*, at 18, Figure 1 inset.
[40] *The State of World Fisheries and Aquaculture 2006*, at 43.
[41] *Ibid.*
[42] Allain, Greenpeace International at 21.
[43] USDA, Foreign Agriculture Statistics, Trade Database, HS-03 fish and seafood, accessed April 2008, available at www.fas.usda.gov/ustrade/.
[44] *See* Food & Water Watch, "Import Alert," July 2007.
[45] European Commission, Health and Consumer Protection Directorate General, "EU Import Conditions for Seafood and Other Fishery Products."
[46] Japan Department of Food Safety, Pharmaceutical and Food Safety Bureau, Ministry of Health, Labor and Welfare, "Statistics of Imported Foods Monitoring for 2005," June 2006.
[47] Food & Water Watch analysis of FDA refusal data received via FOIA.
[48] Another 29.5 percent of the refusals were for other labeling and packaging issues, which can include food safety issues such as unlabeled ingredients that may pose a risk of allergic reaction or packages that are damaged or leaking.
[49] U.S. House Energy and Commerce Committee, Subcommittee on Oversight and Investigations, Staff Statement, "Diminished Capacity: Can the FDA Assure the Safety and Security of the Nation's Food Supply – Part 2," July 17, 2007 at 2-3.
[50] FDA, "Transcript of FDA Press Conference on Seafood Imported from China," at 15.
[51] Linden and Turner, "Aquaculture and Environmental Health in China," at 1.
[52] Huang, Annie, "Chinese Hairy Crab Imports Banned After Tests Find Excess Carcinogens," *Taiwan Journal*, October 27, 2006.

[53] Linden and Turner, "Surf and Turf: Environmental and Food Safety Concerns of China's Aquaculture and Animal Husbandry," at 24.

[54] FDA, "Transcript of FDA Press Conference on Seafood Imported from China," at 4-5.

[55] *Ibid.* at 4.

[56] Nohlgren, Stephen, "Wal-Mart Removes Chinese Catfish Nationwide After Alabama Bans Sales," *St. Petersburg Times*, April 26, 2007.

[57] Becker, Geoffrey S., Congressional Research Service, "Food and Agricultural Imports from China," RL34080, July 17, 2007 at 10.

[58] "FDA Exempts First Chinese Farmed Seafood Producer from Import Restrictions," *Associated Press*, September 19, 2007.

[59] Zhang, Jane, "U.S., Canada Split Over Chinese Shrimp," *Wall Street Journal*, December 15, 2007.

[60] Congressional Research Service at 5.

[61] Rosenthal, Elisabeth, "Europe's Appetite for Seafood Propels Illegal Trade," *New York Times*, January 15, 2008.

[62] Guerin, Bill, "How to Beat U.S. Trade Barriers," *Asia Times Online*, Southeast Asia edition, March 22, 2006, available at *http://www.atimes.com/atimes/Southeast_Asia/HC22Ae01.html.*

[63] Yao, Shunli, China Center for Economic Research, Peking University, Working Paper for the U.S. International Trade Commission, "Chinese Foreign Trade Performance and the China-US Trade: 1994— 2004 – A Geographical Analysis Based on China Customs Statistics," No. 2008-03-A, March 2008 at 6, note 5.

CHAPTER SOURCES

The following chapters have been previously published:

Chapter 1 – This is an edited, excerpted and augmented edition of a United States Congressional Research Service publication, Report Order Code RL34124, dated January 27, 2010.

Chapter 2 – This is an edited, excerpted and augmented edition of a United States Government Accountability Office publication, Report Order Code GAO-09-258, dated February 2009.

Chapter 3 – This is an edited, excerpted and augmented edition of a publication written by David Sefcik for the National Institute of Standards and Technology, dated May 28, 2009.

Chapter 4 - This is an edited, excerpted and augmented edition of a United States Congressional Research Service publication, Report Order Code RS22797, dated January 4, 2010.

Chapter 5 – This is an edited, excerpted and augmented edition of a United States Food and Drug Administration publication, dated June 28, 2007.

Chapter 6 – These remarks were delivered as testimony of Don Kraemer before the U.S.-China Economic and Security Review Commission Hearing on Chinese Seafood, dated April 24, 2008.

Chapter 7 – These remarks were delivered as testimony of E. Spencer Garrett before the U.S.-China Economic and Security Review Commission Hearing on Chinese Seafood, dated April 24, 2008.

Chapter 8 – These remarks were delivered as testimony of Patrick Woodall before the U.S.-China Economic and Security Review Commission Hearing on Chinese Seafood, dated April 24, 2008.

INDEX

D

E

F

G

H

I

J

L

R

S

T

U

V

W